I0486433

THE

FUNDAMENTAL

FORCE

Len Kurzawa

Portrait by Jule Sutton

THE

FUNDAMENTAL

FORCE

HOW THE UNIVERSE WORKS

Len Kurzawa

Order this book online at www.trafford.com
or email orders@trafford.com

Most Trafford titles are also available at major online book retailers.

© Copyright 2018 Len Kurzawa.
All rights reserved. No part of this publication may be reproduced, stored in a retrieval
system, or transmitted, in any form or by any means, electronic, mechanical, photocopying,
recording, or otherwise, without the written prior permission of the author.

Print information available on the last page.

isbn: 978-1-4269-2169-8 (sc)
isbn: 978-1-4269-2170-4 (hc)

Library of Congress Control Number: 2009912362

Because of the dynamic nature of the Internet, any web addresses or links contained in
this book may have changed since publication and may no longer be valid. The views
expressed in this work are solely those of the author and do not necessarily reflect the
views of the publisher, and the publisher hereby disclaims any responsibility for them.

Any people depicted in stock imagery provided by Getty Images are models,
and such images are being used for illustrative purposes only.
Certain stock imagery © Getty Images.

Trafford rev. 04/03/2018

www.trafford.com
North America & international
toll-free: 1 888 232 4444 (USA & Canada)
fax: 812 355 4082

To my sons: Michael, Gregory, Jeffrey, and Kevin

Acknowledgments

Thanks to my son Greg for his encouragement and for the time he spent correcting all my poor punctuation, syntax errors, and bad grammar. His work made this book very smooth reading. And thanks to Brandon Fall for his contribution of those difficult to do diagrams.

Table of Contents

List of Figures

Preface

Me and Einstein.

On April 18th 1955, one of the world's greatest scientists died. Albert Einstein's (1879-1955) life ended at the age of 76, due to a ruptured aorta. I was only ten years old at the time, but I knew quite well who Albert Einstein was, and what he had been doing. I carefully read the newspaper article about his death. I was saddened by the loss of a great scientist, perhaps even more so by the thought that now there would be no more great scientific discoveries about our universe. I thought that perhaps the secrets of the universe would remain hidden forever.

Later that day I sat at the dinner table holding a fork about a foot above the surface. Very slowly I opened my fingers and watched intently as the fork dropped down to the table. It fell, bounced on the hard wooden surface producing a loud clanging sound. I had only to repeat the process a few times before my mother shouted from the kitchen, "Stop making that racket!" In the process of conducting a very serious scientific experiment, I paid no heed. I slowly picked up the fork again. This time I held it only six inches above the table's surface. No, that's still too high. I thought I'd lower it to three inches. That may be too short of distance for a good scientific experiment, but perhaps sufficiently more quiet to escape my mother's notice. I released my grip on the fork, repeating my experiment from a lower height. Then from the corner of my eye I saw my mom standing at the end of the table with squinted eyes. She repeated herself considerably louder than before. "I said stop making that racket!" Oops, I thought, I had better stop or I won't be getting dinner tonight.

Later that evening, I resumed my experiments in a more private location this time dropping coins on my bed, conducting quiet experiments. Besides, I wasn't interested in making noise; I was interested in gravity. I needed to know all about this mysterious force that grips everything in the universe, no matter how far away. The results of my experiments were always the same no matter what I dropped, it always fell down. I dropped forks and spoons; pennies, nickels, dimes, quarters, and half dollars. I dropped them from different heights. The results were always the same. Everything goes down – even in the dark, never up, always down. How does everything know which direction is down?

How does one study something that cannot be examined with their senses? Gravity does its work in silence and beyond our senses. I couldn't see it, I couldn't smell it, and I couldn't taste it. I couldn't feel it either; but I knew it could pull me down. My conclusion, even at an early age, was that gravity always works the same way. Everything is pulled down, rain or shine, night or day, year after year; gravity never gives up; gravity never fails; gravity never sleeps. Gravity is relentless!

That's as far as I got then, but I still wanted to know how gravity works and why. What is this mysterious force that can pull everything down, no matter how big? I knew then that I would forever keep my eyes open, as in my experiments, always looking and hoping to find the hidden secrets of gravity. Little did I know then what amazing discoveries were yet to come.

My search for understanding the universe started with gravity. Once gravity was understood, a light fell on other hidden mysteries of the universe. The shrouds of mystery were pulled away one mystery at a time. Then the larger picture of the universe became clear. It was my dream come true. It is truly an awesome and beautiful vision.

What is presented here is not a compilation of disjointed theories and speculations. What is presented is a logical explanation of how the universe works. Everything described will fit together quite precisely. It is based on solid science from the greatest scientific minds of man; then goes several giant steps beyond. These are bold declarations to make, but the following pages will support this thesis. Several of the greatest unsolved mysteries of science will be explained. Many questions will be answered, many will remain; and, of course, new ones will be raised.

My first objective is to make what is presented clear and easily understood. The second objective is to show the progression of thought, to lead us down a path from one logical conclusion to another, from one discovery to another. In the end we will see the universe is simpler, more beautiful, and even more elegant than we had previously imagined.

Quite a few mathematical equations are used throughout this book. They are important for scientists and others who are technically oriented. These equations are the evidence of the truth within. The mathematics presented can be easily understood. All that's used is high school algebra, a little geometry, and just a touch of trigonometry. Understanding the mathematics is not necessary for those who just want to visualize the magnificence of the universe in which we live.

A quote from a scientist – anonymous.

All my life I have pondered the unfathomable depths of knowledge in search of a better understanding of the mystery of gravity. In all of science it is still something we cannot clearly explain.

Chapter 1

On the Shoulders of Giants

I can see clearly now!

Great Scientists in Search of Knowledge

In this chapter we need to lay down some ground work in preparation for the following chapters. This will help the reader understand the intriguing drama of discovery that is about to unfold. First, we'll need to introduce some of the key players and the roles they played in the process of discovery. The road of discovery is a long one with many obstacles in the way. Great discoveries are made one step at a time; each building on previous ones, and each preparing the way for discoveries yet to be made.

Sir Isaac Newton made the following statement in a letter to fellow physicist Robert Hooke, *"If I have seen a little further it is by standing on the shoulders of Giants."* And so it is, the progress in understanding the nature of the universe is made one step at a time, seeing a little further because of the effort and labor of great scientists that have gone before us. In this chapter, we will be tracing the footsteps of these brilliant scientists who led the way. These giants have laid the foundation upon which we will continue to build.

Because great discoveries are built on those that came before, each paving the way for those to come, it is important to read the chapters in sequence. Latter chapters are built upon premises in prior chapters. Taken out of sequence, the reader will have no basis for understanding or belief, and may dismiss what is being stated as pure conjecture. At the end of this journey, we will have a much different view, and a greater depth of understanding of the universe in which we live. We will see much further, because we will be standing on the shoulders of giants.

Galileo Galilei (1564-1642)

Gravity perhaps came into focus when Galileo Galilei, the Italian mathematician, and physicist invented the telescope. Of course, once he had a telescope he immediately became the world's greatest astronomer. He discovered that the planet Jupiter has orbiting moons, the Sun has spots, and that the Milky Way was composed of dust, and billions of stars. These were only a few of many discoveries. But perhaps of even more importance was his study of gravity. In 1589 he published a treatise on the center of gravity in solids, and conducted experiments which showed that solid objects of different weight will fall at the same speed. He also showed that objects shot into the air will follow the path of a parabola. Through these experiments he demonstrated that there were laws of motion that governed moving bodies. Gravity was at work and gravity was working according to a set of laws.

Johannes Kepler (1571-1630)

Johannes Kepler a German astronomer and mathematician, defined the laws of planetary motion. Kepler built on the astronomical observations meticulously collected by fellow astronomer and Danish nobleman Tycho Brahe (1546-1601). From this data Kepler derived the mathematical formulas that describe the motion of planets around the Sun.

Kepler's Three Laws of Planetary Motion:
1. The Law of Orbits – All planets revolve in elliptical orbits, with the Sun at one of the foci.
2. The Law of Areas – A line connecting a planet and the Sun sweeps out equal areas during equal intervals of time.
3. The Law of Periods – The square of the orbital period of a planet is directly proportional to the cube of the semi-major axis of its orbit.

Although Kepler's laws described the motion of planets around the Sun, the explanation of why the planets obeyed these laws remained a mystery.

Sir Isaac Newton (1642-1727)

Englishman Sir Isaac Newton picked up where Galileo and Kepler left off. He defined the laws of motion according to his law of gravitation. Objects, celestial and terrestrial, strictly obey these laws because, Newton postulated there is a force affecting all matter everywhere. This is the force of gravity. It is gravity that causes the motion of planets around the Sun, and it is gravity that causes objects to fall to the Earth. Newton's laws of motion and gravity explained why the planets obeyed Kepler's laws. It was Newton who now showed that the same laws of motion that govern the planets also govern the motions of objects on Earth.

Newton's Laws of Motion:
1. The Law of Inertia – An object at rest will remain at rest unless acted upon by an external force. An object in motion will remain in motion unless acted upon by an external force.
2. The Law of Acceleration – The rate of change of momentum of a body is proportional to the resultant force acting on the body and is in the same direction as the applied force. (Force = Mass * Acceleration).
3. The Law of Reciprocal Action – All forces occur in pairs, and these two forces are equal in magnitude and opposite in direction. (For every action there is an equal and opposite reaction.)

Newton's law of gravitation states that every particle of mass attracts every other particle of mass by a force along the line connecting them. This force is equal to the product of the two masses multiplied by the gravitational constant, and divided by the square of distance between them.

The gravitational formula as given by Newton is:

Equation 1 – The Law of Gravitation

$$F = \frac{Gm_1m_2}{d^2}$$

Where:
   F is the gravitational force between the two particles of mass
   m_x are the interacting masses

d   is the distance between the two masses

G   is the gravitational constant

Newton's Reservations

Although Newton's understanding and explanation of the laws of motion was a monumentally great accomplishment, he still lamented his inability to identify the cause of gravity. What was the source of this powerful force that could bring the giant bodies inhabiting the universe under its gentle but persuasive sway? What is the source of this amazing power that could hold such massive giants in perfectly controlled orbits at such great distances? Where did this power come from? Newton had no clue. He even refused to offer a hypothesis, believing it would be contrary to sound science. And certainly it would be wrong to play a guessing game with no foundational support.

In the second edition of Principia Mathematica (1713), General Scholium, Newton states:

"I have not yet been able to discover the cause of those properties of gravity from phenomena, and I feign no hypotheses... And to us it is enough that gravity does really exist and act according to the laws which we have explained, and that it abundantly serves to account for all the motions of the celestial bodies..."

The following quote comes from Newton's letter to Richard Bentley Feb. 25th 1693:

"It is inconceivable, that inanimate brute matter should, without the mediation of something else, which is not material, operate upon and affect other matter without mutual contact... That gravity should be innate, inherent, and essential to matter, so that one body may act upon another at a distance, through a vacuum, without the mediation of anything else, by and through which their action and force may be conveyed from one to another, is to me so great an absurdity, that I believe no man who has in philosophical matters a competent faculty of thinking, can ever fall into it."[1]

Newton must have pondered very long and laboriously over the source of gravity. He found no plausible explanation. What could possibly cause one particle of matter to be attracted to another, especially over large distances? He eventually came to the conclusion that only a fool would venture to hypothesize. For two centuries no one could and no one did. Then finally, someone stood up to be that fool with a new idea. That person would be a German scientist named Albert Einstein.

Maxwell's Equations

In 1873, thirty four years before Einstein's theory of relativity, James Clerk Maxwell (1831-1879) a Scottish mathematician and theoretical physicist, wrote his treatise on electricity and magnetism. Maxwell was truly a mathematician of great genius. Working with English chemist and physicist Michael Faraday (1791-1867), Maxwell applied his mathematical skills to Faraday's experiments, which showed a relationship between electricity and magnetism. This resulted in one of the most significant achievements of his short but brilliant career. It was the formulation of a set of equations, now known as Maxwell's equations. These equations integrated and unified electricity and magnetism. So intimate is this relationship that we now refer to them via one name - electromagnetism. Electricity and magnetism are often referred to as two sides of the same coin.

Maxwell also demonstrated that electric and magnetic fields travel through space in the form of waves, which we now simply call electromagnetic waves. But, the big element of surprise was his discovery that light, simple ordinary light, is in fact an electromagnetic wave. Along with this revelation came a rather striking implication. Electromagnetic waves travel at the same speed. If light is an electromagnetic wave, then light also travels through space at the same speed – which is fixed. The speed of light is a constant; and Maxwell's mathematics showed what that speed was.

Maxwell was held in very high esteem by Einstein. It is believed that Einstein kept a picture of Maxwell on his study wall for inspiration; and regarded his work as *the most profound and the most fruitful that physics has experienced since Newton.*" No doubt, Einstein spent much time analyzing the implications of the works of both Newton and Maxwell, two great geniuses that preceded him.

Along Came Michelson and Morley

In the years just before Einstein wrote his thesis on the Theory of Relativity, two scientists: Polish born German-American physicist Albert Michelson and American chemist Edward Morley conducted experiments to find the elusive luminiferous aether which theoretically enabled light to travel through space. Many consider this one of the most famous experiments in the history of science. The aether wind was never found, but more importantly, the experiment proved that light traveled at the same speed regardless of the speed of the light source. Whether the source of light is pointing in the direction of the Earth's motion, or in the opposite direction, it makes no difference. Moving the source of light backward or forward simply has no effect; the speed of light is always the same. For his ingenious ideas on this experiment, Albert Michelson was awarded the Nobel Prize.

Einstein's Dilemma

Now this was something that must have troubled Einstein, although it proved that Maxwell was right about light traveling at a constant speed. But, this experiment gives a new meaning to the word constant. The word constant for the speed of light in empty space really means constant, never changing - no matter what. How could the speed of light be constant in all directions? How can the speed of the light source be totally disregarded? That's bad logic and bad science, not to mention bad karma. How could anyone possibly believe that could be true. It defies logic. If that were true, time and space would have to somehow be distorted. That surely can't be true. Time seems to flow smoothly at the same rate for everyone and space of course would have to be straight. How could space be bent anyway? It's made of nothing, so there's nothing to bend.

Also, if light did travel at a constant speed, then all the laws of motion as defined by Newtonian dynamics wouldn't quite work exactly as defined. Calculations to add and subtract vectors of velocities or forces, among other things would be slightly off when measured at different speeds. Things wouldn't add up exactly right; so calculations of vectors would have to be adjusted for different speeds. This just doesn't seem right. Laws are laws for good reasons. Here is Einstein's dilemma. Einstein realized that Newtonian dynamics as defined by Newton and a constant speed of light as set forth by Maxwell are not compatible.

To rank up there with the great men of science like Newton and Einstein, you have to reinvent the way you see the world[2]. Some long held scientific principles may need to be gutted and redefined. Traditions once firmly held may have to be tossed out with the garbage. Beliefs thought to be true may have to be modified. Who can make those calls? Who is willing to see things in a new light, to view things from a different perspective? It must be someone of genius to make sense of a paradox and give the world a new way to look at things. This is what Albert Einstein was all about.

The Theory of Relativity is Born

No doubt the agony of these issues caused Einstein many sleepless nights. Of course, many great ideas are born in agony. Which will it be: Newtonian dynamics or Maxwell's equations? Newtonian dynamics or Maxwell's equations? Oh no! Another nightmare! These were two of the greatest scientists ever, both men of great genius. Why couldn't they just agree? If only he could get the two of them together to resolve this little discrepancy; but unfortunately, both were dead.

Einstein thought many times about what it would be like to be running along side of a wave of light. What would you see? He labored over this question for quite some time. Then one day he was riding in a streetcar. He looked back to see the clock tower in the town square. Facing forward once again he thought, what would it be like if I were running ahead of the light wave coming from the clock tower? The faster I run the slower the clock would appear to run because it would take the light wave longer to reach me. Let's see, the faster I go, the slower the clock runs, because it takes more time for the light wave to reach me. Now, think about that over and over again. If I were traveling at the speed of light, the clock would appear to be standing still. It wouldn't really be standing still; it would just appear that way to me. People standing in front of the clock tower would see the clock running at the normal speed. With this thought experiment, Einstein concluded that time varies depending on one's frame of reference. But, the speed of light remains constant.

Thus the Theory of Relativity was born. This new concept carries with it many implications that could not have even been imagined until then. Newtonian dynamics had to be viewed in a different light. If fact, the way we viewed the universe had to change. Now, the mass of objects would be different depending on the speed of the mass or person doing the measuring. Time would flow differently for people and things moving at different speeds, or moving in different directions. Now, space is warped. Straight isn't straight anymore. Mass can be changed into energy, and energy can be converted into mass. And mass gets more massive the faster it goes. Who could have even imagined all that?

This doesn't seem to make any sense; it's all very confusing. With Newton's way of thinking, I could see the universe as being well organized, space was straight, time flowed smoothly, mass was mass and it stayed that way. Yes, the universe was clear and clean cut, well organized, and a nice safe place to live. It now seems to me that since Einstein came along, the universe is in total shambles and disarray. Space is warped, time is warped, and so is my psyche. I thought great scientific discoveries were supposed to be simple, beautiful and even elegant. I don't see the elegance here. How can all this be explained?

The Mind of God

Maybe only a genius like Einstein could see the elegance here. And he did. In 1905, when Einstein presented his Theory of Relativity to the world, he had no proof. Even with no proof, he knew this theory would prevail. How could he be so sure, so confident that this view was in fact the view of reality? He knew it was. Because, he knew that if it were not true, the universe would not work correctly. Say what? That's right - The universe would not work correctly. Huh, can you explain that? Well Einstein would say, it's all really very simple. If the speed of light differs depending on the speed of the observer, then all the laws that govern the universe would be different for everyone depending on their speed. The laws of physics would be different everywhere in the universe, because everything in the universe is in motion, and at different speeds. At high speeds, near the speed of light, the differences would be dramatic. At lower speeds the differences would be small but measurable.

This has some very serious implications. Gravitational attraction would vary from place to place: some places weak some strong. In some places galaxies may not be able to form, many may not be able to rotate. The laws of thermodynamics would be different everywhere. Some things would get hot, some not; depends on where you are. In many places, what would have been stars may not be, because they were unable to ignite. Electromagnetism would work differently at different speeds (different places in the universe). Nuclear and molecular interactions would be different from place to place. This means chemistry would be different depending on where you are. Cook books could only be used in the designated areas for which they were written. Your favorite cake receipt wouldn't work in other places in the universe. Everywhere you went in the universe, the rules would change. The universe would not be a very friendly place for travelers. Life would be unsafe at any speed. In fact, life would not be possible in most places.

But if the theory of relativity is true, and the speed of light is constant, then all the laws of physics will be the same everywhere you go in the universe. No matter what speed you travel, everything works the same. The laws of electromagnetism would work the same for everyone everywhere. Gravity – the same everywhere. Thermodynamics – the same. Chemistry – the same. Cooking – the same. You can take your cookbooks with you wherever you go. All recipes would work everywhere, except of course, at high altitudes. The abundance, and variety of life that we see here on the Earth, can prevail everywhere in the universe, always according to the same laws. That's pretty neat!

Maybe this theory of relativity is not such a bad idea after all. It seems to keep everything in the universe simple and under control. The same laws apply everywhere in the universe, that's simple and beautiful too. Okay, maybe even elegant. Perhaps that's why Einstein could see the universe as being superb, a marvel and a masterpiece of creation and a very nice place to live. In the dynamics of the universe, Einstein could see the mind of God.

The Four Known Forces

All physical phenomena that occurs, whether via chemical reactions, radioactivity, nuclear interaction within the stars, motion of bodies, electric or magnetic interactions, could be explained by the four known forces of the universe.

The Four Forces:
Gravity
The Electromagnetic Force
The Weak Nuclear Force
The Strong Nuclear Force

It's through the understanding of these forces that scientists explain everything that happens in the physical world. The nuclear weak and strong forces are the late comers on the scene. They came into our conception to help explain the characteristics of nuclear energy. The concept of a nuclear strong force came into being to explain how the nucleus of an atom could contain multiple positively charged particles, called protons. The weak nuclear force is used to explain the phenomena of radioactivity and fusion reactions in the Sun.

All the physical phenomena we human beings understand today, whether on the Earth or in outer space, are explained by these four forces. These forces are unique because of their different characteristics. They have different strengths, are effective over different ranges of distance, and interact with different types of particles. For example: gravity is the very weakest of all forces. It interacts with all matter over long ranges. Gravity's force is one of only attraction. The electromagnetic force is very strong, interacting over a long range, but only with electrically charged particles. Particles of the same charge are repelled from each other, while particles of opposite charge attract each other. The nuclear strong force is the strongest, but only works at extremely short distances (within the nucleus of an atom), and only interacts with protons and neutrons.

The Grand Unification Theory

Upon close examination of these four forces there doesn't seem to be any way to connect them together. Michael Faraday was the first to conduct experiments demonstrating a connection between electricity and magnetism. Then it was Maxwell and his equations that nailed down their relationships. Faraday's success in connecting two previously unrelated forces spurred him on to more experiments involving gravity. He had a very strong feeling that gravity had some sort of relationship with electricity and magnetism. Even though his experiments never showed a connection, he never faltered in his belief. He died believing the connection existed.

A breakthrough was made in 1979 when Stephen Weinberg, Abdus Salam, and Sheldon Lee Glashow received the Nobel prize for their discoveries connecting the weak nuclear force to the electromagnetic force. This was referred to as the Electroweak Unification. Some few scientists still continue on in the effort to unify all forces. The theory to unite all forces is often referred to as the Grand Unification Theory, or GUT.

Einstein also believed that somehow these forces were interrelated. But how? What could be their connection? He spent years contemplating, but never came to a resolution. Four forces in the universe are used to explain all physical phenomena. That's not bad. But, shouldn't they be related? Einstein thought so, but could never develop with a comprehensive theorem. This was his primary objective: to know and understand how the forces of the universe were related to each other. He died, as did Faraday, believing that all the forces of the universe were somehow connected.

However, there is one very important clue Einstein left for us that should not be dismissed as simply an interesting observation. Einstein believed that mass distorted space. He said, " The geometrical properties of space ... are determined by matter."[3] Perhaps this is more eloquently stated by American physicist John Wheeler, when he said, "Matter tells space how to curve, and curved space tells matter how to move." It is the distortion of space that bents light waves and causes planets to follow a curved path around the Sun. Einstein's discourse on the structure of space being the cause of gravity was not sufficient to captivate the imagination of scientists or induce them to continue down the same path. Quantum physics has taken precedence. His theory on gravity has stood still since the time he presented it – until now.

Getting Back on Track

The world of science has gone off to quantum and string theory, dark matter and energy, the big bang, and of course black holes. No major breakthroughs have been made and no great changes have occurred in the way we view the universe during the past several decades. Are we off track?

In the following chapters of this book, we will be going back to where Einstein left off. We've been stuck in a rut much too long. Let's pick up the pieces, get back on track, and then begin moving forward once again. We will reinvent the way we see the universe. Be ready and willing to toss out old beliefs that don't work. Others may have to be gutted and redefined. We are going to take a different perspective and look at things in a different light. When we're finished, we're going to see the universe in a totally different way, and with more clarity than ever before.

Here's a brief glimpse of what we're about to see. We will see that there is one fundamental force that powers the universe, and from it all other's are derived. We will see that gravity is not a force, but an effect. We will see how mass contracts space and what that means. We'll understand the motion of bodies in space including the revolution of planets and the rotation of galaxies. We'll finally understand why we have tides, and why we have high tides on both sides of the Earth at the same time. We'll see where matter comes from and how it is created. Once we know where matter comes from, we'll be able to see how the universe was created, and what it will be like in the future.

For those scientists wondering about the origin of the fine-structure constant, its explained somewhere in this book. For those scientists looking for Dark Matter, it's explained here. For those scientists trying to solve the Pioneer Anomaly, the solution is here. And last, but not least, a logical explanation of The Grand Unification Theory, how all the known forces are connected together.

Chapter 2

Kepler's Laws

Planets in motion abide by laws.

Kepler's Law of Orbits

Kepler's laws are key to understanding the motion of bodies in space, and will be referred to frequently. These laws are the foundation upon which we can build to help explain some of the unsolved mysteries of the universe.

Kepler's first law, The Law of Orbits, states that all planets move in elliptical orbits with the Sun located at one focus. In all ellipsis there are two foci. They are on opposite sides of the center and are equidistant from the center. Eccentricity is the measure of how far off center the foci are. The ratio of the distance of the focus from the center divided by the size of the major axis is the measure of eccentricity. This value is a ratio and can be anywhere from 0, (a perfect circle) to smaller than 1 (extremely elliptical). Eccentricities of 1 or larger, would not apply to ellipsis, but to parabolas. Eccentricities of 1 or larger would apply to comets that may take a trip around the Sun, but never return. Comets that do not return follow a parabolic path. Planets in orbit follow an elliptical path.

Table 1 – Kepler's Law of Orbits, shows the eccentricity for each of the planets in our solar system. The table shows a very low eccentricity level for all the planets, except Mercury. Of course this means that Mercury has the most elliptical path. The rest of the planets have orbits that are very close to being circular. Although Pluto has a highly eccentric orbit, it has lost its status as a planet, so has been dropped from this table.

To get an idea of the shape of the elliptical path of any planet, use the column showing the Degree of Tilt. Take a perfectly round object like a Frisbee, or dinner plate. Hold it out away from the body with outstretched arms so as to see the full circle of the plate. Then tilt it very slightly to the angle specified in the table. The elliptical shape the eye perceives would be the shape of the planet's orbital path.

Let's use the Earth as an example. Hold a dinner plate with outstretched arms so you are seeing it full circle. Then tilt it 1 degree. See the difference. No, neither do I. If the orbit of the Earth, or any of the planets, save Mercury, were drawn on a piece of paper, the human eye would not be able to distinguish it from a perfect circle. So, when scientists tell us the orbits of the spheres in space are elliptical, just nod your head. They are just being their scientific selves, with their scientific preciseness. As for me, when I say the planets are traveling in circular orbits, please excuse me, I mean nearly circular orbits.

Try the same experiment for Venus and Neptune. For Venus the tilt would only be 1/3 of a degree; for Neptune, ½ of a degree. The most eccentric planet is Mercury, for which the tilt is 11.8 degrees. And it still looks like a circle to me.

Table 1 – Kepler's Law of Orbits

Planet	Radius in millions of Km	Eccentricity	Degree of tilt	X Size in millions of Km	X Size in millions of Miles
Mercury	57.91	0.20563	11.8	11.908	7.3996
Venus	108.21	0.00677	.38	.73258	.45522
Earth	149.60	0.01671	.95	2.4998	1.5533
Mars	227.94	0.09341	5.3	21.291	13.230
Jupiter	778.41	0.04839	2.7	37.667	23.406
Saturn	1426.73	0.05415	3.1	77.257	48.007
Uranus	2870.97	0.04716	2.7	135.39	84.131
Neptune	4498.25	0.00858	.49	38.595	23.987

In this table, the columns labeled "X Size" list the number of kilometers or miles the focus is off center. This can be used to calculate the aphelion (point of orbit farthest from the Sun) and perihelion (point of orbit closest to the Sun) of the planet. Using the Earth as an example again, the eccentric size is about 1.5 million miles. Add that to the radius of the Earth's orbit, 93 million miles, to get the aphelion – 94.5 million miles. Subtract 1.5 million miles from the Earth's orbit to get the perihelion – 91.5 million miles. The eccentric sizes may seem like large distances, but they are not when compared to the distance of the planet from the Sun. The eccentricity shows that all the planets except for Mercury are fairly close to having circular orbits.

A Planet's eccentricity can be quickly calculated by dividing the X Size by the Radius. For example, Mercury's focus if off center by 11.908 million kilometers, and its radius is 57.91 million kilometers. Dividing 11.908 by 57.91 gives us the eccentricity of 0.20563. The angle of tilt can be obtained by the inverse sine of the eccentricity ($\sin^{-1}$ e). For example: $\sin^{-1}$ 0.20563 = 11.8 degrees.

Kepler's Law of Areas

Now let's look at Kepler's second law – The Law of Areas. Kepler showed us that as a planet revolves around the Sun it will sweep out equal areas of space during a specified time frame. Figure 1 illustrates this with the elliptical orbit of the planet Mercury.

Figure 1 – Kepler's Law of Areas

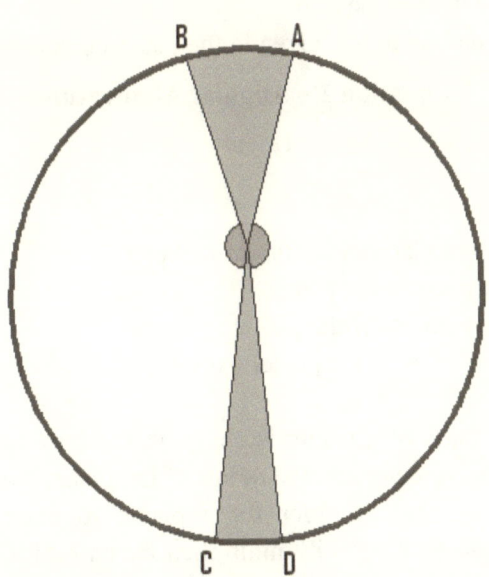

Picture the planet Mercury, as it revolves around the Sun, from high above the plane of Mercury's orbit. Mercury takes 88 days to make its revolution around the Sun. After every 11 days imagine a line between Mercury and the Sun. After 8 iterations of 11 days we would be able to envision 8 areas of space. Because the orbit of Mercury is highly elliptical, all areas would appear as triangles of different shapes and sizes; yet all 8 areas would be equal in size.

The ellipse pictures Mercury's orbit with the Sun at one of the foci. The time it takes Mercury to travel from points A to B is the same amount of time it takes to travel from points C to D. This is because Mercury's velocity is greater at its perihelion than at its aphelion. Kepler discovered that the two shaded areas are equal to each other in size.

Although this appears to have no practical value to us, it does illustrate the geometric precision of planetary motion and the preservation of angular momentum. As a planet revolves around the Sun it always maintains the same angular momentum. The conservation of angular momentum is also illustrated in Figure 1. Mercury's velocity is greater when traveling from point A to point B because it is closer to the Sun. Mercury's velocity is slower when traveling from point C to point D because it is further from the Sun.

The angular momentum is given in the equation below:

Equation 2 – Angular Momentum

$$L = mvr$$

Where:

   L   is the angular momentum of a planet
   m   is the mass of the planet
   v   is the velocity of the planet
   r   is the radius of the planet's orbit

Because the mass of a planet is constant, the only two variables that can change are the velocity and the radius of the planet's orbit. We can see from this equation that in order for a planet to maintain its angular momentum, the planet's velocity multiplied by its radius must always be the same. So, as the radius increases, the velocity must decrease. This explains why the planets that are farther away from the Sun move along at a slower pace. The closer a planet is to the Sun, the faster it travels.

This law is also a prelude to a new law of planetary motion that is coming in a later chapter. This also leads to Kepler's third law which follows in the next section.

Kepler's Law of Periods

Kepler's Law of Periods states that the square of the period of any planet is proportional to the cube of the semi major axis of its orbit. This relationship is expressed in the equation shown below. The distance from the Sun cubed (d^3) divided by the period of time it takes a planet to go around the Sun squared (t^2), is the same for all the planets in the solar system. When the distances are expressed in Astronomical Units and the period of rotation is expressed in Earth years then the distance cubed would be equal to the period squared - $d^3 = t^2$.

The following is an equation derived from Kepler's third law that illustrates the harmony of planetary revolution:

Equation 3 – Kepler's Law of Periods

$$m = \frac{4\pi^2 d^3}{Gt^2}$$

Where:

m   is the mass of the Sun in kilograms.
d   is the distance a planet is from the Sun in meters.
t   is the time it takes a planet to make one revolution around the Sun, in seconds.
G   is the gravitational constant.

Let's do the math and calculate the mass of the Sun from the Earth's distance and time of revolution:

$$m = 1.989 \times 10^{30} = \frac{4\pi^2}{6.6742 \times 10^{-11}} \frac{(1.496 \times 10^{11})^3}{(3.1557 \times 10^7)^2}$$

The Fundamental Force

The results of this calculation is the same for all planets, no matter how massive the planet. Notice there is no mass specified in this equation for the planets. It's from the relationship of the planet's time of revolution and distance from the Sun that the Sun's mass can be attained. Remember, Kepler derived his laws of planetary motion through observation, without knowing the mass of any of the planets. Table 2 – Kepler's Law of Periods shows the results of this calculation for each of the planets.

Table 2 – Kepler's Law of Periods

Planet	Distance in meters	Orbital Period in seconds	Mass of Sun in kilograms
Mercury	5.791×10^{10}	7.6005×10^{6}	1.989×10^{30}
Venus	1.0821×10^{11}	1.9414×10^{7}	1.989×10^{30}
Earth	1.4960×10^{11}	3.1557×10^{7}	1.989×10^{30}
Mars	2.2794×10^{11}	5.9355×10^{7}	1.989×10^{30}
Jupiter	7.7841×10^{11}	3.7436×10^{8}	1.990×10^{30}
Saturn	1.42673×10^{12}	9.2929×10^{8}	1.989×10^{30}
Uranus	2.87097×10^{12}	2.6514×10^{9}	1.991×10^{30}
Neptune	4.49825×10^{12}	5.2004×10^{9}	1.990×10^{30}
Pluto	5.90637×10^{12}	7.8238×10^{9}	1.990×10^{30}

In Table 3 – Kepler's Law of Periods, which is a continuation of Table 2, we can see that there is a relationship between the distance a planet is from the Sun cubed and the time it takes a planet to orbit the Sun squared. The first three columns in Table 3 show each of the planets, their distance from the Sun in Astronomical Units (AU), and the time it takes to orbit the Sun in years. An Astronomical Unit is defined as the mean distance of the Earth to the Sun.

The two columns on the right show the distance from the Sun in AUs cubed, and the orbital period in years squared. Notice the last two columns are equal. However there is some discrepancy for the planets more distant from the Sun. This discrepancy may possibly be attributed to the imprecise measurements of distance and orbital period. Then again, there may be another reason.

Table 3 – Kepler's Law of Periods

Planet	Distance in AUs	Orbital Period in years	Distance Cubed	Period Squared
Mercury	0.387	0.2408467	0.06	0.06
Venus	0.723	0.6151972	0.38	0.38
Earth	1.000	1.000	1.00	1.00
Mars	1.524	1.8808476	3.54	3.54
Jupiter	5.203	11.862615	140.88	140.72
Saturn	9.537	29.447498	867.45	867.16
Uranus	19.191	84.016846	7,068.23	7,058.83
Neptune	30.069	164.79132	27,186.63	27,156.18
Pluto	39.482	247.92065	61,544.20	61,464.65

These two columns are plotted Figure 2 – Kepler's Law of Periods. The straight line connecting the planets depicts the relationship between a planet's distance from the Sun and its period of orbit. The farther away a planet is from the Sun, the slower its velocity, and the longer it takes to make a complete cycle.

Figure 2 – Kepler's Law of Periods

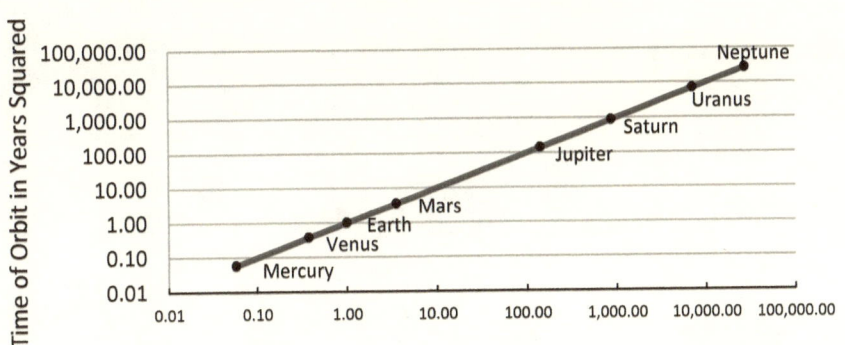

Chapter 3

Gravity

"Gravity – it's not just a good idea, it's the Law!" – NASA slide

Two Views of Gravity

In this chapter we will take a look at two views of gravity. First let's take a look at how gravity works in the classical Newtonian view. Then we'll pick up the concept of gravity as introduced by Einstein. Einstein introduced a new concept of gravity as being related to the structure of space. But, he did not elaborate on this new concept. He didn't show us how gravity really works or give us a complete explanation of what gravity is. What he did, was leave us with clues and a direction to go. Now is the time to pick up where he left off.

Newton's Law of Gravitation

First, let's take a look at how gravity works in the world of Newtonian dynamics. Then let's see how gravity works in Einstein's world of structured space.

The gravitational formula as given by Newton is:

Equation 4 – The Law of Gravity

$$F = \frac{Gm_1m_2}{d^2}$$

Where:

F is the gravitational force between two masses
G is the gravitational constant
m is the mass of the objects.
d is the distance between the two masses

Newton's law of gravitation states that, "Every particle in the universe attracts every other particle with a force that is directly proportional to the product of their masses and inversely proportional to the square of the distance between them" The gravitational constant is the strength of the gravitational field between two objects. The gravitational constant is often referred to as a constant of proportionality.

Measuring Gravity

The strength of a gravitational field is defined by a constant of proportionality denoted by the capitol letter "G". The gravitational constant cannot be derived by calculations from other constants. There are no fundamental constants related to the gravitational constant that can be used to validate it's correctness. This constant is implicitly measured by various apparatus specifically invented for that purpose.

In 1798 British chemist and physicist Henry Cavendish (1731-1810), was the first to invent such an apparatus, by which he could measure the gravitational constant. He conducted an experiment to measure the force of gravity between two very heavy lead balls. Two 350 pound lead balls were suspended by wires to a torsion bar balance. The instruments were finely tuned to detect even the slightest tension between the two lead balls. This was the first accurate measurement of the gravitational constant. Once this constant became known, calculations could be made to determine the masses of the Earth, Sun, moon, and planets, as well as other heavenly bodies.

The accuracy of the gravitational constant is best described in an abstract of an article by George T. Gillies.[4]

Abstract. Improvements in our knowledge of the absolute value of the Newtonian gravitational constant, G, have come very slowly over the years. Most other constants of nature are known (and some even predictable) to parts per billion, or parts per million at worst. However, G stands mysteriously alone, its history being that of a quantity which is extremely difficult to measure and which remains virtually isolated from the theoretical structure of the rest of physics. Several attempts aimed at changing this situation are now underway, but the most recent experimental results have once again produced conflicting values of G and, in spite of some progress and much interest, there remains to date no universally accepted way of predicting its absolute value.
...

An article by Robert Kritzer entitled; "The Gravitational Constant"[5], describes several interesting methods used to measure the gravitational constant. The article also describes the difficulties involved in attempting to get an accurate value for the gravitational constant. The article shows some of the methods used and the values obtained. The values vary but are not far from the values Henry Cavendish originally obtained in his experiments.

The current value of the gravitational constant recommended by CODATA[6] is:

$$G = \frac{6.67428 \pm 0.0010 \times 10^{-11} \ m^3}{kg * sec^2}$$

The value listed on the NASA website[7] is:

$$G = \frac{6.67259 \pm 0.00030 \times 10^{-11} \ m^3}{kg * sec^2}$$

The Gravitational Force Between the Earth and Moon

We can now use Newton's equation (Equation 4) to calculate the force between any two objects. Let's use the Earth and the moon for an example. The mass of the Earth is 5.974×10^{24} kilograms; and the mass of the moon is 7.348×10^{22} kilograms. The distance from the Earth to the moon is approximately 3.844×10^8 meters. Utilizing these values in Equation 4 we get:

$$F = \frac{Gm_1 m_2}{d^2} = \frac{6.6742 \times 10^{-11} * 5.974 \times 10^{24} * 7.348 \times 10^{22}}{(3.844 \times 10^8)^2}$$

$$F = 1.9827 \times 10^{20} \ Newtons$$

The force of gravity between the Earth and the moon is 1.9827×10^{20} newtons. This may not mean much if you don't normally work with newtons; but, this is a very powerful force. A Newton is defined as the amount of force required to accelerate a mass of 1 kilogram by 1 meter per second squared. As can be seen from the very large numbers above, an extremely massive force is required to keep the moon in orbit around the Earth. This formula applies to all the planets, moons, stars, and galaxies in the universe. As the masses get larger, so the force of gravity must increase to keep everything in its proper orbit. If we could sum up the total force required to keep the universe operating as is, the number of newtons required would be too large to comprehend.

An Alternate View - Acceleration

Instead of calculating the force of gravity, we can look at the situation a little differently. Another equation Newton provided us is referred to as the Law of Acceleration. That is Force = Mass * Acceleration. This is another equation for force. Einstein suggested that the force of gravity as defined by Newton's Law of Gravity and the force of acceleration on mass as defined by Newtons second law are essentially equivalent. This is referred to as The Equivalence Principle. This equation for force is shown in Equation 5:

Equation 5 – The Law of Acceleration

$$F = ma$$

Where:

F   is the force being applied
m   is the mass being accelerated
a   is the rate of acceleration

Making a substitution of mass and acceleration (ma) for force in Equation 4, we can see that the force of acceleration on a body is equivalent to the force of gravity on the body.

$$m_2 a = \frac{G m_1 m_2}{d^2}$$

Let's assume that m_1 is the mass of the Earth and m_2 is the mass of the moon. Dividing the mass of the moon from each side, we now have an equation, shown below, which gives us the acceleration of the moon toward the Earth due to the force of gravity. By looking at the acceleration caused by gravity instead of force, we can get a different perspective on gravity. This is an important point. We will be using Equation 6 over and over again to help illustrate the nature of gravity.

Equation 6 – Acceleration

$$a = \frac{Gm}{d^2}$$

Where:

a   is the rate of acceleration of the attracted mass due to the gravitational force of the attracting mass (m). In many cases a lower case 'g' is used to denote the acceleration caused by gravity.

G   is the gravitational constant.

m   is the mass of the object doing the attracting.

d   is the distance between the two masses.

Using this equation we can determine the acceleration of the moon toward the Earth.

$$a = \frac{6.6742 \times 10^{-11} * 5.974 \times 10^{24}}{(3.844 \times 10^8)^2} = 0.002698346 \text{ meters/sec}^2$$

Converting the acceleration from meters to millimeters we have:

$$a = 2.698346 \text{ millimeters / sec}^2 = .10628 \text{ inches / sec}^2$$

The calculation above shows that the moon falls toward the Earth about 2.69 millimeters every second, or about one tenth of an inch every second. That's doesn't seem like very much, but it's just enough to keep the moon in orbit around the Earth. The moon travels about 1020 meters per second at a tangent that is perpendicular to the Earth. Every second the gravitational force of the Earth pulls the moon toward it just one tenth of an inch. Of course if it did, then the moon would still be the same distance from the Earth. In one month, the moon would be back where it started, relative to the Earth.

The word acceleration as used by scientists means a change of speed or a change of direction. Note that the speed of the moon traveling around the Earth does not change. Acceleration here refers to the moon's constant change of direction toward the Earth. This is a very small change; about one tenth of an inch for every kilometer. This is just enough change to keep the moon in orbit around the Earth. This is why the scientists say that the moon is always falling toward the Earth. Always falling, but never getting any closer.

When we calculated the force of gravity between the Earth and the moon we obtained the total force of both combined. In the calculation for acceleration we are only seeing one side of the picture. The result we have obtained is for the acceleration of the moon toward the Earth. We obtained this result because we used the mass of the Earth in the equation. If we wanted to see the acceleration of the Earth toward the moon we would have used the mass of the moon instead of the Earth. Since the mass of the Earth is over 81 times greater than the moon; the moon's effect is very small, so we need not be concerned about it here.

Notice also that in Equation 6 there is only one mass to be specified. We did not need the mass of the moon to calculate its acceleration toward the Earth. Remember Galileo's experiment (dropping objects from a tower), which showed that solid objects of different weight will fall at the same speed. So the size or weight of the moon makes no difference. It will fall toward the Earth according to the mass of the Earth.

Einstein's View of Gravity

What was just described is nothing new, that was the Newtonian view. That's the way gravity is viewed in most of our high school text books. Now we'll take a look at gravity as related to the structure of space. Einstein used the terms, "structure of space" and "geometry of space."

Popular terms now used to describe the structure, or geometry, of space are "warped space" and "curved space." But just what do they mean?

Actually, I don't like using the term "warped space." It has a bad connotation, like something is wrong, distorted, bent out of shape, twisted or disfigured. However, the meaning of the word "warp" as related to space comes right out of Webster's Dictionary with the following meaning: "to turn or incline from a straight, true, or proper course; to deviate." This definition turns out to be an accurate description of what actually happens.

The term "curved space" is often used because objects traveling through curved space would tend to travel in a curved path. Perhaps this term would be better to use. In the next few paragraphs I'll use them both. Then we'll try to find a more descriptive and meaningful term.

The straight, true, or proper course of any object moving through space is a straight line. In fact, this follows from Newton's law of inertia. An object in motion will remain in motion, continuing in the same direction, unless acted upon by an external force. Previously we had considered the force of gravity as being that external force acting on objects traveling though space. Of course, when Newton derived his laws of motion he had no concept of warped or curved space. Now, Einstein has changed all that, and that puts a twist on Newton's law of inertia. Now we have to say that an object traveling through space will travel in a straight line unless acted upon by an external force, or the object travels through warped space.

Consider an object traveling through space. Unhindered by any force and traveling through "straight space" (usually referred to as flat space), the object would travel at the same rate of speed and in the same direction, in a straight line, forever. However, when our traveling object approaches a massive object, it is entering into warped space, and the path of our traveling object turns toward the massive object. The path the object takes is curved. The degree of curvature would depend on the amount of curvature in space. So we can say that space is curved near massive objects. The amount of curvature of space would be determined by the mass of the object being approached and the distance from it.

It would take a considerable amount of force to change the direction of a massive body like the moon for instance. But when traveling through warped space there is no force being applied. Warped space does not apply a force. An object traveling through space would simple follow the course laid out by curved space as if it were a straight line. The natural, proper and true course of objects traveling through curved space will be a curved course. This is the true and proper course because that is the straightest course in curved space. No force need be applied.

Contracting Space

What we've just described as warped space may still be confusing. The reason for the confusion is our lack of understanding of what exactly is warped space, and how it becomes warped. The term "warped space" still carries a negative connotation because we do not yet understand it. Instead of thinking in terms of warped space, let's think of contracted space. Think of contracted space in the sense that space is shortened, shrunken, or reduced in size. Yes, that's it – contracted space. Now this concept of space being contracted should not be too difficult a concept to accept. Just think of space as being elastic; easily contracted or expanded. Since space is empty, a volume of empty space can be easily reduced in size. That should be simple enough, but it has implications.

Einstein tells us: "the geometrical properties of space are not independent, but they are determined by matter."[8] Mass is the culprit that causes space to contract. Mass, all mass, has the property of contracting the space surrounding it. The more massive the object, the greater the degree of contraction. The contraction of space is directly proportional to the mass of the object. We are simply changing the way we see things. Previously we would have said: every particle of mass attracts every other particle of mass with a force we call gravity. Now let's say: every particle of mass contracts the space around it.

Now here's a new concept that follows from contracted space. Objects in or traveling through contracting space tend to turn toward the source of contraction. Objects in contracting space appear to be attracted to each other, and they are. But not by a force; the attraction is the effect of the contraction of space. This is gravity! Gravity is not a *force*, it is an *effect*. Gravity is the effect of contracting space. No force is applied to the objects, they are simply following the natural and proper course of any object traveling through contracting space.

All around the Earth, on every side, in every direction, space is contracting toward the Earth. The amount of contraction is directly proportional to the mass of the object. Also, the amount of contraction is inversely proportional to the distance from the mass squared. If the distance is doubled, the contraction of space is one fourth. If the distance is tripled, the contraction is only one ninth. This is referred to as the inverse square law. This inverse square law is nothing new. This is the way gravitational fields work. Whether considered a force, or an effect, gravity still works the same way and the inverse square law still applies. The only difference is that now we're applying it to the contraction of space as opposed to a force of gravity.

Let's formalize these statements into two laws of contracting space:
1. Every particle of mass contracts the space around it. The amount of contraction of space is directly proportional to the mass of the object causing the contraction and inversely proportional to the distance from the mass squared.
2. Objects in or traveling through contracting space accelerate toward the source of contraction.

Statement number 2 is worded in a way to cover objects in motion and objects not moving relative to the object contracting the surrounding space. Objects in motion would turn toward the source of contraction. The "turning toward" is an acceleration. Objects not in motion would accelerate directly toward the source of contraction. For now, let's temporarily accept these two statements as possibly new laws of physics. More explanation that shows these statements to be true will come later.

Space Contractions

Let's look at the relationship between the Sun and the Earth in terms of contracting space. We'll use Equation 6 again, this time to calculate the rate of acceleration of the Earth toward the Sun. Let's just say the cause of the Earth falling toward the Sun is not because of a gravitational force, but because the space around the Sun is contracting. Equation 6 remains the same, what has changed is our interpretation of what it means.

The Sun's mass is 1.989×10^{30} kilograms and the distance between the Earth and Sun is 1.496×10^{11} meters. Let's calculate.

$$a = \frac{Gm}{d^2} = \frac{6.6742 \times 10^{-11} * 1.989 \times 10^{30}}{(1.4960 \times 10^{11})^2} = 5.9316 \times 10^{-3} \ m/s^2$$

$$a = 5.9316 \text{ millimeters} / sec^2 = 0.2337 \text{ inches} / sec^2$$

Previously we would have said that the acceleration of the Earth toward the Sun is caused by the gravitational force of the Sun. Now let's say the acceleration of the Earth toward the Sun is caused by the contracting space around the Sun. How much did space contract? The Earth is a distance of 1.496×10^{11} meters from the Sun. At that distance space contracted about 6 millimeters, or about ¼ of an inch. The Earth accelerates toward the Sun by that amount every second.

We could also say that the Earth falls toward the Sun about ¼ of an inch every second? One would think, even though this is a very small amount, that eventually the Earth would finally fall into the Sun. But that never happens. It never happens because in our case "accelerate" means a change of direction. The change of direction is toward the Sun.

This is such an important concept to understand so we should examine it more closely. Let's analyze what happens. The Earth is traveling through space at about 29.785 kilometers per second at an angle perpendicular to the Sun. During that second, space between the Sun and the Earth contracted by ¼ of an inch. The Earth followed the space contraction in a curved path. This is also considered to be an acceleration toward the Sun of ¼ of an inch. If there had been no contraction of space, and the Earth traveled in a straight line, in one second it would have moved ¼ of an inch away from the Sun. But, fortunately, space did contract around the Sun and the Earth followed in a curved path, falling ¼ of an inch closer to the Sun. This was a perfect exchange of space: ¼ of an inch lost, and ¼ of an inch gained; a perfect balance which keeps the planets in nearly circular orbits. So, the Earth is continually falling into the Sun, but never getting any closer.

This is the most wonderful and amazing phenomenon that could possibly be imagined. This doesn't apply just to the Earth, it applies to everything. Everything is falling. All moons and planets in the solar system are falling around the Sun. All stars in the Milky Way are falling around its center. All stars in all galaxies are falling. All of this falling is what maintains their orbits. And all of this falling is caused by contracting space. No force need be applied to anything to keep it in orbit. All objects in space, everything is in perpetual motion, with no force whatsoever being applied. Everything is just falling.

Equation 6 – Acceleration, has been used to calculate the contraction of space at the radius of the planets orbiting the Sun. The rate of contraction is listed in Table 4 in the "Contraction" column. The mass of the Sun is about 1.989×10^{30} kilograms; which is more than 99.8% of the total mass of the whole solar system. The radius of each planet's orbit is given in millions of kilometers; and the contracted space size is given in millimeters per second squared.

Table 4 – Space Contractions

Planet	Mass in Kilograms	Radius in meters	Contraction in millimeters
Mercury	3.3020×10^{23}	5.791×10^{10}	39.5857
Venus	4.8690×10^{24}	1.0821×10^{11}	11.3372
Earth	5.9742×10^{24}	1.4960×10^{11}	5.9316
Mars	6.4191×10^{23}	2.27936×10^{11}	2.5551
Jupiter	1.8987×10^{27}	7.78412×10^{11}	0.2190
Saturn	5.6851×10^{26}	1.42673×10^{12}	0.0652
Uranus	8.6849×10^{25}	2.87097×10^{12}	0.0161
Neptune	1.0244×10^{26}	4.49825×10^{12}	0.0065
Pluto	1.3000×10^{22}	5.90637×10^{12}	3.805×10^{-6}

From this table we can see that the Earth is at a distance of 149.6 million kilometers (93 million miles) from the Sun. At this distance, space is contracted by about 5.9 millimeters per second. That's not very much – about ¼ of an inch. But this is sufficient to keep the Earth in its orbit around the Sun. The amount of contraction decreases with distance. The farther away a planet is from the Sun, the less contraction of space at that distance. Once again, the amount of space contraction decreases with the square of the distance.

Neptune is one of the giant planets at the edge of the solar system. Neptune has a mass of 1.0244×10^{26} kilograms and is 4,498 million kilometers from the Sun. At that distance, the contraction of space is only a tiny .0065 millimeters per second. But this is all that's necessary to keep Neptune from escaping into outer space.

Let's take a look at Jupiter, the most massive planet in our solar system. Jupiter weighs in at a whopping 1.8987×10^{27} kilograms, and is 317 times more massive than the Earth. Contracting space at Jupiter's location is only .219 millimeters per second; a mere fifth of a millimeter. That's smaller than the width of the period that ends this sentence. But that's all that's necessary to keep this massive planet in orbit around the Sun.

Note also that the mass of the planet makes no difference in its orbit. The mass of the planet, although listed in the table, was not used to calculate the rate of contraction. If the orbits of the Earth and Jupiter where exchanged, there would be no difference. The Earth would follow the same path that Jupiter did; and Jupiter would do likewise, following the same path that the Earth did. This is because the path laid out by contracting space would be the same for any planet at that location. The mass of the planet is not a factor in determining the path a planet takes; neither is a "force" of gravity acting on the planet. The amount of contracting space is what determines the path.

Gravity Recap

What has been described in this chapter are the very basic laws and principles of gravity. Einstein's view of gravity via structured space has been expanded to gain an understanding of space contraction. Our understanding of space contraction is still not complete and will be enhanced in the next few chapters.

Chapter 4

Gravitational Considerations

Gravity is not a force, it's an effect.

Things to Consider

So far we have looked at gravity in two different ways. First we looked at the Newtonian view of gravity as a force, and then we looked at Einstein's view of gravity as an effect of structured space. The mathematics of Newton's formula remains the same regardless of which view is taken; only the interpretation changes. No proof has been presented yet that would make one view more preferable than the other. Some considerations here will show that viewing gravity as an effect of contracting space is both simple and elegant. The balance should then tip heavily in favor of this new view of contracting space. Here are some things to consider.

The Gravitational Constant

Let's take another look at the Gravitational Constant.

$$G = \frac{6.6742 \times 10^{-11} \ m^3}{kg * sec^2}$$

Of interest now is not the value of the Gravitational Constant itself but the units of measurement.

$$\frac{meters^3}{kilogram * second^2}$$

As mentioned earlier, the gravitational constant is often referred to as a constant of proportionality. The units of measurement were carefully chosen so that calculations using this constant would yield the correct units of measurement. But the units of measurement also give meaning to the constant as well. Meters cubed are the units of measurement for a volume. Cubic meters per kilogram is referred to as a specific volume. For now we can interpret the Gravitational Constant to be a measurement of a specific volume of space being contracted by a kilogram of mass. It can be viewed as a rate of attraction or contraction.

Expended Energy

Let's assume gravity is a force acting on all mass. In the previous chapter we calculated the force required to keep the moon in orbit around the Earth. Multiplying that force by the distance the moon moves gives the amount of energy expended in Joules. A tremendous amount of energy would have to be expended every second to keep the moon in orbit. This would be an ongoing continuous expenditure of energy. This has been going on for billions of years, yet the energy never runs out, is never exhausted, and never diminishes. Gravity never gives up, never fails. But from where does this inexhaustible supply of energy come?

And that's just the moon, what about all the other planets in our solar system? Why does the Sun never tire of keeping numerous and massive planets under control? What about all the stars in the Milky Way? It would take an incredible amount of force and energy to maintain the revolution of a galaxy. Where could all this energy come from?

If we view gravity not as a force, but as an effect of contracting space, then there is no expenditure of energy whatsoever. All bodies traveling through space move without effort in their natural and proper course, without force. No energy is necessary and no force is needed to maintain the revolution of moons, planets or galaxies.

Managing Multiple Objects

If gravity is a force, consider the difficulty the Sun has in dealing with multiple objects in orbit. The Sun must apply just the right amount of force on each object in its vicinity to maintain their orbits. The amount of force needed is directly proportional to the mass of the object in orbit. The force needed to keep the Earth in orbit around the Sun is much greater than the force required to keep the moon in its orbit around the Sun. So the Sun needs to apply exactly the right amount of force on each object to keep everything in balance. The force of attraction applied to the Earth must be greater than the force applied to the moon, even though they are the same distance from the Sun.

How does the Sun know how much force to apply to the Earth and the moon respectively to keep them in orbit? Going back to Newton's Law, we could say that the force required is directly proportional to the product of their masses. If we calculated the force required using Newton's equation we would see that the force required to keep the Earth in orbit is 81 times greater than that necessary for the moon. This makes sense as the Earth is 81 times more massive than the moon. So it would appear that bodies in orbit would have to contribute to the force necessary in order to keep themselves in orbit.

But consider this: we already know from the equation for acceleration that the masses of the objects are not required in the equation to calculate their acceleration toward the Sun. Both the Earth and the moon accelerate toward the Sun at the very same rate even though they have a significant difference in mass. Their acceleration is independent of their mass. But their acceleration toward the Sun is directly proportional to the mass of the Sun. So, whatever the cause of the attraction, the mass of the attracted object is of no consequence.

If gravity was an effect of contracting space, none of this would matter. The amount of space contraction would be the same for both the Earth and moon so they would both maintain their orbits in unison. All planets and their moons would simply follow the course laid out by the structure of space they're in. Once again, no force is necessary.

Gravitational Tug-of-War

There is another consideration in managing multiple objects by a force of gravity. None of the objects being held by a gravitational force can diminish the force applied to other objects. Let's illustrate this with a game of tug-of-war. Let's apply the rules of this game to the solar system. None of the moons or planets in the solar system can diminish the gravitational force the Sun applies to any of the other moons or planets. The fact that the moon is in close proximity to the Earth cannot diminish the force of attraction that the Sun applies to the Earth; and vice versa.

Suppose a massive object the size of the Earth came visiting one day from outer space. The Sun would have to apply an additional force to attract this new visitor and still maintain the same hold on all the other moons and planets it already keeps in orbit. Where did this additional force come from? There would seem to be an endless supply of power that must come and go as necessary. The point here is that the force of gravity cannot diminish or be divided between multiple objects.

Once again, with gravitational attraction caused by contracting space there simply is no problem. Multiple bodies and visiting objects would all travel according to their "straightest" path in curved space. No need for additional forces to be conjured up. If fact, there is no need for any force at all.

The Missing String

Science teachers often use the analogy of a ball on a string being swung around overhead to represent the moon in revolution around the Earth. The tension on the string represents gravity holding the moon in orbit. The problem with this analogy is that there is no string. There is no physical connection between the Earth and the moon. There is nothing between the Earth and moon but empty space. So where's the string? As we have already seen, there would be incredibly powerful forces at work between the massive bodies in space if gravity is a force, and a powerful connection is necessary for the transmission of powerful forces. There is no sign of any such connection. The search still goes on for gravitons, a hypothetical particle, used to transmit the gravitational force.

Consider the problems gravitons have to deal with. Every particle of mass attracts every other particle with a strength that diminishes with the distance between them squared. Nothing can be put between the two particles that will diminish that attraction except space. Nothing but space!

When a total eclipse of the moon occurs, the Earth comes directly between the Sun and the moon. This does not block nor diminish in any way the gravitational attraction between the Sun and the moon. This means the connection between the Sun and the moon still remains exactly as it was even with the Earth directly in between. The gravitons would have to "know" what body they are connecting with. Some gravitons would be required to pass right through the Earth to reach the moon, while others "know" to connect with the Earth. This is just not plausible.

Just one other point to add to this issue of a connection between masses: the connection must work both ways; in both directions; as we have seen in a previous section on Managing Multiple Objects. The object being attracted must also contribute to its gravitational force. So the connection must be bidirectional.

If gravity is the results of contracting space, then no connection is necessary. The motion of the moon will continue to respond to the contraction of space no matter what intervenes. There is no need for a string. Gravitons haven't been found because gravitons aren't there.

Gravity Cannot Be Blocked

Both electric and magnetic attraction are easily blocked. Although these forces are much stronger than gravity, they can be cut off by intervening material. Their fields also obey the general inverse square law, so they theoretically have the same reach as gravity. However, since these fields are easily blocked, they are effectively only short range forces.

The gravitational field cannot be blocked – by anything. Gravity is a very weak force with a very long range. It knows no bounds. When an eclipse of the moon occurs, the Earth is positioned directly between the Sun and the moon. This does not block the gravitational attraction the Sun has on the moon. Gravity is relentless. It cannot be interrupted, redirected, refocused, retracted, relaxed, or regulated. Gravity works, all the time, every time – without fail. Why is that so? We can understand the answer to that question only if we change our view of what gravity is.

We have always thought of gravity as a force between particles of mass. Newton said: "every particle in the universe attracts every other particle with a force that is directly proportional to the product of their masses …" It's hard to understand how all particles of mass throughout the universe can be connected to all other particles of mass. That's entirely too many connections – requiring too many gravitons. This is just not palatable.

We need another way to state the case for gravity that still maintains the same effect. Let's say that every particle of mass contracts the space around it. Let's also say that the rate of contraction is directly proportional to the size of mass doing the contraction. And also that the contraction of space is a continuous process. This is what mass does – contract space.

Let's also assume that mass will follow the path of contracting space. It would then appear that mass is being attracted to the object that is contracting space. This is essentially the same as accelerating toward the object doing the contracting. Let's also assume that all mass is contracting space at a rate inversely proportional to the square of their distances. What we have just described, is gravity without the force. The effect is still the same; objects of mass are still attracted to each other. However, there is no force involved. Fields of force can be blocked, contraction of space cannot. Therefore, gravity cannot be blocked.

One question to naturally arise is what happens to all that contracted space. Can space just keep contracting forever? This is an exceedingly important issue that will require more discussion. The answer, presented in a later chapter, – will be astounding. For now, let's accept these assumptions to be true, qualified with the understanding that they will be thoroughly explained later.

The Travels of Light

Einstein predicted that gravity would bend light. He was proven correct when experiments by English astronomer Sir Arthur Eddington in 1919 showed that light waves passing near the Sun were in fact bent. What bent the light waves? Was it the force of gravity or the effect of contracting space? Consider this - light waves have no mass. Newton's law of gravity states that every particle of mass attracts every other particle of mass. How can light waves, which have no mass, be affected by gravity? The light waves must have bent because of the gravitational effect of contracting space. Even light waves follow the natural, true and proper course determined by curved space, which is a curved path. Einstein predicted that – he knew something! In a forth coming chapter we will see what Einstein saw.

Chapter 5

Space Contractions

It's with contractions that babies and great ideas are born.

Untangling the Mystery of Gravity

In a mystery novel as more and more information is acquired, the identity of the antagonist comes into focus with more clarity. It's like doing a jigsaw puzzle. With each new piece added, the picture becomes clearer. The suspense mounts in anticipation of the final pieces, the final clues, that give a solution to the mystery and the identity of the culprit.

The solution to a complex mystery never comes easily. Facts are acquired one at a time. They are analyzed and scrutinized to the utmost degree. But even then sometimes the facts are incomplete or misinterpreted. Assumptions are made; but assumptions are often wrong. Trails are followed down wrong paths. When a dead end is reached, the puzzled but cunning sleuth knows he must backtrack. Where was the wrong turn taken? What were the false assumptions? Which facts were interpreted incorrectly? What critical information is still missing?

So it is the same in attempting to solve the complex mysteries of the universe. It's always one piece at a time, and the pieces must all fit together to make a clear picture. When a piece is missing then information is missing. When information is missing there are questions that remain unanswered. In previous chapters we made assumptions and left questions unanswered.

In this chapter we will add pieces to the puzzle and answer some of the questions that had been put on hold. This will give us a greater depth of understanding and a clearer picture of how gravity really works. Some questions could not be explained in preceding chapters because the answers would introduce new questions which would only add confusion. A foundation had to be built which would allow new concepts to be introduced at the right time. That time is now.

I have been promoting the idea that gravity is the contraction of space, and the contraction of space is caused by matter. The questions carried over from an earlier chapter are:

- How are human beings attracted to the Earth?
- What happens to all that contracted space?
- Can space continually contract forever?

The answer to these questions are yet to come.

The Contraction of Space

In this section we will clarify our understanding of gravity by getting a better grasp on what is meant by the term "contraction of space". My previous suggestion was to think of contraction in the sense of shorting, shrinking or reducing the size of space. That is a good definition, general enough to cover various circumstances. But now a more detailed explanation is needed, and other words of greater precision used for different circumstances. The different circumstances are celestial and terrestrial mechanics.

Gravity works the same in space as it does on the Earth. We have been using the word "contraction" and the phrase "contraction of space" in describing how the moon is attracted to the Earth, and how the planets are attracted to the Sun, etc. In the case of celestial mechanics, our terminology remains the same. A more detailed explanation of space contraction in the celestial world will follow shortly. This will include a new law of physics. But first we will discuss the contraction of space on the Earth.

In the world of terrestrial mechanics, what happens when contracting space encounters the object of mass that is attracting it? We could say that space is contracted into the Earth. We could also say that matter absorbs space. What is being stated here is that the space that is being contracted toward the Earth is actually being absorbed into the Earth. Of course, this is true of all bodies in space. Bodies of matter are absorbing the surrounding space!

For now, when contracting space encounters the attracting matter, let's say that space is absorbed into the matter attracting it. In this case the word "absorbed" is better used in place of the word "contracted". The word "contracted" is not being replaced, but in this case the word "absorbed" gives a clearer picture. A detailed explanation of space absorption will follow immediately in the next section.

Here is a brief overview of a way to look at space contraction. To illustrate we'll use the Earth as our example. The Earth is absorbing space. So we can say that the space surrounding the Earth is absorbed into the Earth. The space just absorbed into the Earth needs to be replenished. So the space above that space is contracted toward the Earth. And the space above that does likewise; etc. We may have to consider that space is a very real entity.

The assumption of space being absorbed by matter may not seem plausible right now. But, bear with this assumption and continue to read on objectively. The evidence will be forthcoming! Perhaps now is a good time to summarize in my own words the statements Isaac Newton made to his good friend Richard Bentley. It is inconceivable that an inanimate body of matter can attract other bodies through empty space with no physical connection. It is an absurdity that defies logic. And so it is! What is about to follow is a logical explanation of how that happens.

The Absorption of Space

For now, let's move forward with the assumption that space is absorbed into the Earth. The rate of absorption is calculated with the same equation we've been using to determine the acceleration of objects attracted to the Earth by gravity. Let's use that equation to once again calculate the acceleration of objects as they fall toward the surface of the Earth.

$$a = \frac{Gm}{d^2} = \frac{6.6742 \times 10^{-11} * 5.974 \times 10^{24}}{(6.371 \times 10^6)^2}$$

$$a = 9.8231 \quad \text{Meters per second per second.}$$

This calculation tells us that the acceleration of objects near the surface of the Earth will cause the objects to fall, toward the Earth at about 9.8 meters per second squared. In the explanation of the gravitational attraction of planets toward the Sun we stated that the gravitational attraction is due to the contraction of space between the Sun and the planets. For the Earth, we said that the space between the Earth and Sun contracted by 5.9316 millimeters, or about ¼ of an inch.

The explanation is exactly the same for all of us here on the Earth. We are all being attracted to the Earth at a rate of about 9.8 meters per second every second. We are attracted to the Earth for the same reason that the moon is attracted to the Earth and planets are attracted to the Sun. Space between us and the Earth, or the space we are in, is contracting toward the Earth, and being absorbed into the Earth. We do not fall into the Earth because the powerful electromagnetic fields of the Earth and our bodies, at the molecular level, are much stronger than the gravitational attraction on us.

Before continuing on with this explanation we also need to understand that space can flow through matter as if it were not even there, just as matter can flow through empty space as if there were nothing there. It's saying the same thing! Remember, matter is mostly empty space anyway. The flow of space though is not like the wind. The effects of the wind can be clearly seen and felt. There is no discernible effects of space traveling through matter, or matter through space, other than the effect of gravity. The Earth is traveling through space at almost 67,000 miles per hour and we don't feel any sensation. We cannot feel or sense the movement of space. Neither can we feel or sense the radio waves, or neutrinos, passing through our bodies. This is happening all the time, yet is unperceived by our senses. Realize that all bodies in space can travel through space uninhibited. Space does not block the motion of objects flowing through it. Now we'll continue on with the explanation of space absorption.

We had previously stated that objects in contracting space move toward the source of contraction. So, our bodies, being made of matter, tend to move along with the contraction of space we're in. If we are just standing on the Earth, once the space we were occupying is absorbed into the Earth, we are still left standing there as if nothing had happened. We never feel the flow of space through us. It's beyond our senses.

Allow me to repeat this in a different way. Let's say were standing here on the Earth, occupying space, and just minding our own business. We are standing in a space that extends from the surface of the Earth, to 9.8 meters above the Earth. That would be about 32 feet from the ground up. In one quick second those 9.8 meters of space above the ground are absorbed into the Earth. The space, (not the air, or anything else contained within the space), is being absorbed into the Earth every second.

The volume of space absorbed can be calculated with the well known volume equation shown below. Calculate the volume of the Earth, then calculate the volume of space using the radius of the Earth plus 9.8231 meters. Subtract the former volume from the latter. The mean radius of the Earth is 6.371×10^6 meters.

Equation 7 – Volume of a Sphere

$$v = \frac{4\pi r^3}{3}$$

The volume of space absorbed by the Earth every second is equal to: $v_2 - v_1$.

$$v_2 = \frac{4\pi (6.371 \times 10^6 + 9.8231)^3}{3} \qquad v_1 = \frac{4\pi (6.371 \times 10^6)^3}{3}$$

$v_2 - v_1 = 5.0104 \times 10^{15}$ cubic meters.

The volume of space absorbed by the Earth every second is 5.0104×10^{15} cubic meters. Keep this number in mind as we shall see it again.

Flowing Space

Now that's all very interesting, but also of interest is what happens to the space above the space that was just absorbed. That space comes down toward the Earth to fill in for the area just vacated by the space just absorbed. Of course, the space above that space does likewise; and on and on it goes. The key here is that even empty space abhors a vacuum. That sounds very familiar, like we've heard that before. Empty space fills the space vacated by empty space. That statement may sound preposterous, but it is not. Remember, empty space is not really empty. It may be void of matter but filled with energy.

We could calculate the volume of the space above the space absorbed and find that it is the same volume as the space which was absorbed. That seems reasonable because that's the volume of space that needs to be filled. We could do the same calculations for the space above that; but let's not bother. Let's take a short cut. This calculation requires a computer or a good calculator, because of the accuracy required. Let's do the same calculation only this time we're shooting for the moon. We've already determined, in the chapter on gravity, the gravitational acceleration the Earth has on the moon, .002698346 meters/sec^2. The mean distance between the Earth and moon is 3.844×10^8 meters.

Let's do the calculation:

$$V_2 = \frac{4\pi \, (3.844 \times 10^8 + .002698346)^3}{3} \qquad V_1 = \frac{4\pi \, (3.844 \times 10^8)^3}{3}$$

$$V_2 - V_1 = 5.0104 \times 10^{15} \text{ cubic meters}$$

The volume of space absorbed by the Earth every second is 5.0104×10^{15} cubic meters. That is exactly the same volume that is contracting towards the Earth every second at the distance of the moon. That is what attracts the moon towards the Earth. It's space contraction. This is gravity!

What are these calculations telling us? The gravitational acceleration of the moon toward the Earth according to the laws of gravitation by Isaac Newton (acceleration equation in the chapter on Gravity), is 0.002698346 meters/sec^2. Call it 2.69 millimeters, or about 1/10th of an inch. Envision this: a volume of space, 1/10th of an inch thick, completely surrounding the Earth, at the distance of the moon. The volume of that space is 5.0104×10^{15} cubic meters. Exactly the same as what is absorbed by the Earth every second.

Visualize what is happening. All matter absorbs space. The Earth, composed of matter, is absorbing space. The amount of space absorbed can easily be calculated. The space absorbed is replace by the layer of space immediately above it and adjacent to it. That space is also replaced in the same manner. Each layer of space is thinner than the one below it because the radius and circumference are continually expanding. But the volume of space being attracted toward the Earth always remains the same. There is a constant flow of space towards the Earth. This flow of space creates a smooth flow in the fabric of spacetime that continues throughout the universe.

Gravity Waves

When Newton proposed his theory on gravity he thought that the gravitational effect on bodies would act instantaneously across the space separating the interacting bodies. Since then, Einstein has shown us that the speed limit is the speed of light. It would be prudent to stay with Einstein's conclusions. It takes time for space from upper regions to fill in the lower regions. I would suggest that this happens at the speed of light because there in nothing in the empty space but energy. So the thought here is that space flows smoothly toward its destination – mass. But the gravitational effects of the flow of space, ripple through space at the speed of light.

Consider the time it takes upper layers of space to fill in the vacated space below it. At the surface of the Earth space is being absorbed into the Earth. Each successive layer of space above fills in the vacated layer immediately below it. Because this flow does not happen instantaneously, a ripple in the fabric of space time occurs. It would take one second for the space 0.3 billion meters away to begin to "feel" the effect of the vacated space below it. This ripple in the fabric of space can be thought of as a gravity wave. That is all that will be mentioned of gravity waves, but there is a chapter on The Flow of Space coming next.

The Law of Volumes

We have already discussed all three of Kepler's laws of planetary motion. The Law of Volumes is one that Kepler missed. What will now be described is a new law of physics and is somewhat similar to Kepler's Law of Areas, and could be referred to as Kepler's 4th Law. Kepler was a brilliant mathematician; and it would take a brilliant mathematician to even imagine that the areas of a planet's elliptical orbit would be of equal size. Kepler's calculations were all done the hard way, with no calculator or computer to aid in his work.

Now with modern technology we can make complex calculation in seconds to analyze complex problems. So with the aid of technology we can see something he missed. We will now see that the volume of space absorbed into a massive object in one second, is the same volume of space flowing toward that object at any given distance. This also illustrates, and validates mathematically, the contraction of space as the source of gravitational attraction. The Law of Volumes explains how gravity works.

The calculation used for the volume of space can be quite cumbersome. Fortunately there is a simpler way. Once we know the gravitational acceleration of a planet toward the Sun, we can readily calculate the volume flow with the following equation.

Equation 8 – The Law of Volumes

$$v = 4\pi r^2 a$$

Where:
    v   is the volume of contracted space.
    r   is the radius of an object's orbit.
    a   is the gravitational acceleration of the object.

The area of a sphere can be obtained from the expression – $4\pi r^2$. Multiply that area by the thickness of that area to get a volume of space. In this case that thickness is the acceleration toward the Sun. That multiplication would give the expression – $4\pi r^2 a$. That volume is the same volume of space that is being contracted toward the Sun every second.

First let's validate this equation with the Earth and moon; then we'll use it for the Sun and planets. The radius of the moon's orbit is $3.844e^8$ meters. The gravitational acceleration of the moon towards the Earth is 0.002698346 meters/sec^2. The following calculation will give us the volume of space at the moon's orbital radius that is contracting toward the Earth every second. Then we'll compare this volume with the volume of space being absorbed into the Earth.

$$v = 4\pi * (3.844\text{x}10^8)^2 * 0.002698346 = 5.0104\text{x}10^{15} \text{ m}^3$$

Let's do the same calculation only this time we'll use the Earth's radius and the gravitational acceleration at the Earth's surface. This calculation will give us the volume of space being absorbed into the Earth every second.

$$v = 4\pi * (6.371 \times 10^6)^2 * 9.8231 = 5.0104 \times 10^{15} \text{ m}^3$$

These results match what has already been calculated in previous sections. These calculations show that the volume of space that is absorbed into the Earth, and the volume of space contracting toward the Earth, at the distance of the moon, is exactly the same. The following figure attempts to help visualize this relationship. The problem is that it's a two dimensional picture attempting to relate three dimensional volumes. Of course the diagram is highly out of proportion so that the visualization can be seen.

Look at Figure 3 – The Law of Volumes, and envision the inner circle at the center as being the Earth. Also envision the Earth being surrounded by a volume of space 9.8231 meters thick at the surface of the Earth encircling the whole sphere of the Earth. The volume of that space is 5.0104×10^{15} cubic meters.

Figure 3 – The Law of Volumes

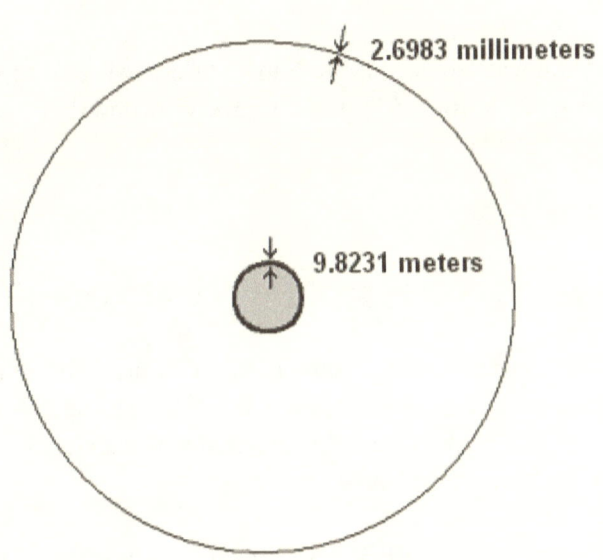

2.6983 millimeters

9.8231 meters

The outer circle represents the orbit of the moon. The acceleration of the moon toward the Earth is 2.6983 millimeters/sec^2. Assume the outer circle is only 2.6983 millimeters thick. Envision that outer circle encircling the whole Earth in all three dimensions. The volume of space 2.6983 millimeters thick, encircling the whole Earth is also 5.0104x10^{15} cubic meters. This is the same volume of space encircling the whole Earth at the surface of the Earth. As the Earth absorbs space, the same amount of space absorbed is also contracted toward the Earth from the surrounding area.

The only relationship we've analyzed so far has been the Earth and moon. Now let's do the same for the solar system to get a broader perspective. First let's use the same equation to calculate the volume of space absorbed by the Sun every second. The mass of the Sun is 1.989x10^{30} kilograms and its radius is about 6.960x10^8 meters. Let's calculate the acceleration due to gravity at the Sun's surface.

$$a = \frac{Gm}{d^2} = \frac{6.6742 \times 10^{-11} * 1.989 \times 10^{30}}{(6.960 \times 10^8)^2}$$

$$a = 274.041 \qquad \text{Meters per second squared.}$$

Using the acceleration at the Sun's surface we can now calculate the volume of space absorbed by the Sun every second. Let's use Equation 8 to make that calculation:

$$v = 4\pi r^2 a$$

$$v = 4\pi * (6.960 \times 10^8)^2 * 274.041 = 1.6682 \times 10^{21} \text{ m}^3$$

Now let's calculate the volume of space at the Earth's radius from the Sun, that is being contracted toward the Sun. The radius of the Earth's orbit is 1.496e^{11} meters and the gravitational acceleration at the Earth caused by the Sun is 5.9316x10^{-3} m/s^2.

$$v = 4\pi * (1.496 \times 10^{11})^2 * 5.9316 \times 10^{-3} = 1.6682 \times 10^{21} \text{ m}^3$$

The volume of space absorbed by the Sun is the same as the volume of space being contracted toward the Sun at the Earth's orbital radius.

Table 5 gives the radius of each planet's orbit, and the rate of acceleration (contraction) toward the Sun. The table shows the calculated results for each planet, giving the volume of space flowing toward the Sun at the specified orbital radius.

Table 5 – Volume of Contracted Space

Planet	Radius in meters	Contraction in meters	$4\pi r^2 a$ in cubic meters
Mercury	5.791×10^{10}	39.5857×10^{-3}	1.6682×10^{21}
Venus	1.0821×10^{11}	11.3372×10^{-3}	1.6682×10^{21}
Earth	1.4960×10^{11}	5.9316×10^{-3}	1.6682×10^{21}
Mars	2.2794×10^{11}	2.5551×10^{-3}	1.6682×10^{21}
Jupiter	7.7841×10^{11}	2.1909×10^{-4}	1.6682×10^{21}
Saturn	1.42673×10^{12}	6.5216×10^{-5}	1.6681×10^{21}
Uranus	2.87097×10^{12}	1.6106×10^{-5}	1.6682×10^{21}
Neptune	4.49825×10^{12}	6.5606×10^{-6}	1.6681×10^{21}

Look at the results of the calculations for volume using the expression, $4\pi r^2 a$, in the last column of Table 5. These calculations show that the volume of space that is absorbed into the Sun and the volume of space flowing toward the Sun is the same at any distance. The flowing of space explains gravity. Objects in space are accelerated in the direction of the space flow. The degree of acceleration depends on the mass and distance from the object absorbing space.

Notice that there is some discrepancy with distance from the Sun, just as in Kepler's Law of Periods. The radius of the planets orbits at the farther distances from the Sun are not precise. We may possibly attribute this to the inaccuracy of human measurements, or possibly another reason we are not yet aware of.

Abiding by Newton's Law

Let's see if what we have just described is in keeping with Newton's Law of gravitation. Newton's Law states: "Every particle in the universe attracts every other particle with a force that is directly proportional to the product of their masses and inversely proportional to the square of the distance between them"

It would seem highly unlikely that all particles in the universe can have a connection with all other particles in the universe. Consider every single proton, neutron, and electron having a connection to all others – in the universe. Because this is what appears to be happening and the laws work, we've accepted this explanation for hundreds of years. There was no other explanation until Einstein. Still, we didn't understand the implications of what Einstein said. But now we have a logical explanation.

Every particle in the universe does attract every other particle in the universe, but not with any specific connection to them. Every particle of matter in the universe absorbs space. This absorption of space happens in the immediate vicinity surrounding each particle of matter. In other words, each and every particle of matter absorbs the space surrounding it. That causes a flow in space (contraction of space, etc.) which causes the attraction of every other particle of mass in the universe. This works all the time, every time, without fail, and nothing can stop it or block it. It works through all objects and it reaches throughout the vast regions of the entire universe.

This simple concept now explains how every particle of mass affects every other particle of mass throughout the universe. What Newton thought to be an inconceivable attraction of bodies through empty space, can now be explained because of a rather simple phenomenon. What was once an absurdity that defied logic, can now be explained by simple logic.

Newton also said that the force of attraction between objects is "inversely proportional to the square of the distance between them". Yes, this is absolutely true, and now we know why. The diminution of gravitational attraction with the increase of distance as described above correlates with the decreasing width in the volume of space. In fact, it is exactly the same thing. The width of the volume of space contracted is also the magnitude of the gravitational attraction. The diminishing width of volume of space with distance is exactly the same as the Law of Inverse Squares. There is no need to go through the mathematical proof of this because it is exactly the same procedure in both cases. Once again, the gravitational attraction due to spatial contraction diminishes according to the Law of Inverse Squares.

What has just been described not only conforms to Newton's Law but explains it. We have taken a whole different view of gravity that now gives a logical explanation to the phenomena of gravity. The evidence is clear. The evidence has been presented and conclusions can be drawn. But, it's not time yet. Remaining questions still need to be answered. More evidence is yet to come. The whole picture is not yet complete.

Chapter 6

The Flow of Space

Go with the flow!

Free Falling Objects

Galileo Galilei, the famous Italian mathematician and physicist of the 16th century, conducted an interesting experiment. He dropped objects of different weights off the top of the Leaning Tower of Pisa. His objective was to see if the heavier objects would fall faster than the lighter ones, as Aristotle has claimed. He discovered in the simple experiment, that objects of different weights fall at the same speed. This same experiment was conducted in 1971 by Apollo 15 astronaut David Scott on the moon where there is no air resistance. The American astronaut dropped a hammer and a feather from the same height at the same time. Both objects fell at the same speed and hit the ground simultaneously.[9]

In this chapter we will see how the flow of space affects objects that are not in orbit. After that, we will take a close look at the flow of space from the moon to the Earth. In this section we will see how the flow of space causes objects to fall. To help us understand this let's visualize Galileo's experiment all over again. Much of what is being presented will be a review of what we had learned in our high school science class. But, now it's being presented from a different perspective.

The Leaning Tower of Pisa is only about 56 meters (184 feet) high. So, instead of using the Leaning Tower of Pisa let's imagine something much higher. Let's assume we have a leaning tower 490 meters (1607 feet) high. The tip of the antenna on the Empire State Building is 449 meters (1472 feet) high. But, this is close enough to give us a visual idea of what we want for our hypothetical experiment.

In free fall the distance an object falls is given by Equation 9:

Equation 9 – Distance of Free Fall

$$d = \frac{at^2}{2}$$

Where:
- d is the distance an object falls in meters
- a is the rate of acceleration in meters/second squared
- t is the time of travel in seconds

Let's assume we drop a ball from the top of a tower 490 meters high. Let's also assume there is no wind or air resistance to affect the rate of falling. The acceleration rate increases very slightly each second as the object approaches the Earth. For our experiment the slight increase over a short range is insignificant near the Earth's surface, so we'll consider the acceleration rate to be constant, near the Earth's surface. We hold a ball in one hand and a stop watch in the other. At the same time we let go of the ball and we start the timer on the stop watch. Let's also assume we have some highly sophisticated technical equipment to monitor every second of this experiment. Once released, the ball immediately starts its plunge downward.

What happens after the first second? We know that the rate of acceleration for free falling objects at the Earth's surface is about 9.8 meters per second squared. After the first second the ball has fallen only 4.9 meters. What happened? At the completion of the first second the ball is falling at a rate of 9.8 meters per second but it only fell half that distance. That's because when we first released the ball it was falling at a rate of 0 meters per second. It was standing still relative to the Earth. Then it took a whole second for the ball to get up to the speed of 9.8 meters per second; which is the rate of acceleration. So the 4.9 meters the ball actually traveled is the average of the starting and ending velocities times the time of travel.

Table 6 is used to correlate the related information of free falling objects. The first column gives the lapsed time in seconds. The second column gives the rate of acceleration. Note that the rate of acceleration remains the same; this is okay for short distances. The rate of acceleration actually increases each second but the increase is very small and need not be of concern in this experiment. The third column shows the velocity at which the ball is traveling at the completion of each second. The fourth column shows the actual distance traveled in that second. And the fifth column shows the cumulative distance traveled as the seconds lapse.

Note that there is no column for the mass of the falling object. As long as the mass is significantly smaller than the mass of the Earth, we need not be concerned. Whatever the mass of the balls we drop, they will all fall at the same speed. With no air resistance, a ping-pong ball will fall just as fast as a bowling ball.

The *Velocity* column in Table 6 shows the velocity at which the ball is falling after each second. This is calculated by multiplying the rate of acceleration (9.8 m/s) by the number of seconds traveled. Notice that the velocity after each second is equal to the previous velocity plus the rate of acceleration. So each second the velocity of the ball increases by the rate of acceleration. After 10 seconds the velocity has reached 98.0 m/s.

Table 6 – Free Fall

Time in Seconds	Rate of acceleration	Velocity each second	Distance each second	Total Distance
0	0	0	0	0
1	9.8	9.8	4.9	4.9
2	9.8	19.6	14.7	19.6
3	9.8	29.4	24.5	44.1
4	9.8	39.2	34.3	78.4
5	9.8	49.0	44.1	122.5
6	9.8	58.8	53.9	176.4
7	9.8	68.6	63.7	240.1
8	9.8	78.4	73.5	313.6
9	9.8	88.2	83.3	396.9
10	9.8	98.0	93.1	490.0
		490	490	

Analyzing Free Fall

We can calculate the distance the ball travels each second by adding the starting velocity to the ending velocity and dividing by 2. This gives us the average velocity the ball was traveling for that second; which is also the distance traveled for that second. For example: the distance the ball traveled in the third second is equal to (19.6+29.4) / 2 = 24.5 meters. But there is another way to determine the distance traveled. The distance traveled in any one second is equal to the distance traveled in the previous second plus the rate of acceleration. Let's calculate the third second in the "Distance each second" column again using distance instead of velocity: $14.7 + 9.8 = 24.5$ meters. This is how the "Distance each second" column in Table 6 was calculated. So here again we see something similar to what we had previously seen with the increase in velocity. The distance the ball traveled each second is equal to the previous distance plus the rate of acceleration.

The velocity of the ball after each second is equal to the previous velocity plus the rate of acceleration. The rate of acceleration is cumulative for both distance traveled and velocity.

Let's use Table 6 to try to understand what is happening. Two very important observations can be made from this table:

1. The column entitled *Velocity each second* shows that the velocity of free falling objects increases every second by the acceleration rate, in this case 9.8 m/s^2.
2. The column entitled *Distance each second* shows that the distance traveled by a falling object also increases every second by the acceleration rate, also in this case 9.8 meters.

Close examination of Table 6 shows the validity of the statements above:

1. The average velocity (98/2), multiplied by the number of seconds gives the total distance of the fall: $49.0 * 10 = 490$ meters.
2. The sum of all the distances traveled each second is equal to 490 meters.
3. Equation 9 from above was used to calculate the values in the *Total Distance* column and the total distance the ball dropped was 490 meters.

From this data we can see that the speed and distance traveled by free falling objects is determined by the rate of acceleration, which is determined by contracting space. Objects in or moving through contracting space move toward the source of contraction (2nd Law of Contracting Space). Objects will move toward the source of contraction at the rate of acceleration as described in Newton's Law. The velocity of the object will increase every second by the rate of space contraction. The distance the object travels every second will also increase by the rate of space contraction. The acceleration of objects is the effect of gravity which is the result of contracting space.

The velocity of objects in free fall increases every second. This shows that they are not carried along in the space they occupy. On the surface of the Earth space is absorbed into the Earth at a rate of 9.8 meters per second. The flow of space is smooth and steady. Over short distances the motion of space can be thought of as flowing at a constant rate. The effect that the flow of space has on objects is an acceleration toward the source of contraction. The flow of space is toward the contracting object. Every second that space is absorbed into the Earth, that same rate of contraction is added to the velocity of all free falling bodies. This is what gives all the free falling bodies acceleration; normally specified in meters per second squared.

Objects have momentum, space does not. The rate of velocity is cumulative in falling physical bodies. Thus, momentum increases as objects fall. At the end of the first second of free fall, the falling objects are traveling at the same velocity as that of contracting space. After the first second, free falling objects will fall faster than the flow of contracting space.

All of this is true for any kind of ball dropped from our tower. Golf ball, baseball, football, soccer ball, or volleyball, it's all the same. This is also true of human beings. We all fall the same way the balls fall.

Figure 4 illustrates the speed and distance which the ball in our experiment falls.

The Fundamental Force

Figure 4 – Free Fall

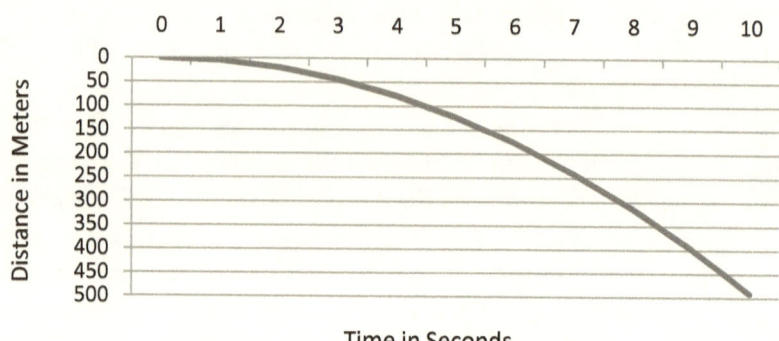

Notice that after 4.5 seconds the ball had only traveled a distance of about 100 meters. After another 4.5 seconds, a total of 9 seconds, it had traveled about 400 meters. The velocity and distance traveled increase with time according to the acceleration rate.

Moon Fall

Now let's take a look at what would happen if the moon would fall into the Earth. This could only happen if the moon was not in orbit around the Earth. So for this thought experiment, let's pretend the moon is in a stationary position in relationship to the Earth. Let's assume the distance from the Earth to the moon, from center to center is 3844×10^8 meters. Let's start the timer on our stop watch and observe the moon fall. The first second goes by. At that distance, space is flowing toward the Earth at 2.6983×10^{-3} meters per second. That's only about 2.7 millimeters per second, or about $1/10^{th}$ of an inch per second. The acceleration of the moon toward the Earth is also $1/10^{th}$ of an inch per second. The moon's speed after one second is also $1/10^{th}$ of an inch per second. The distance traveled by the moon is $1/20^{th}$ of an inch.

The next second passes. We can calculate the speed by simply adding the acceleration rate to the previous speed. That would be $2.6983 + 2.6983 = 5.3966$ millimeters per second. After 2 seconds the moon is traveling at the speed of $1/5^{th}$ of an inch per second. Every subsequent second we can add the rate of acceleration at that distance to get the current speed. The start is very slow, but the speed increases every second. Table 7 tells the rest of the story.

Table 7 – Moon Fall

Hours	Distance in 10 hours	Distance Traveled	Remaining Distance	Velocity in meters/sec
10	1,750,723	1,750,723	382,649,276	97
20	5,284,411	7,035,134	377,364,865	196
30	10,669,211	15,953,622	368,446,377	299
40	18,016,468	28,685,679	355,714,320	408
50	27,495,283	45,511,751	338,888,248	527
60	39,357,834	66,853,117	317,546,882	660
70	53,988,229	93,346,063	291,053,936	815
80	72,002,355	125,990,584	258,409,415	1,005
90	94,487,604	166,489,959	217,910,040	1,258
100	123,730,251	218,217,855	166,182,144	1,650
110	166,815,678	290,545,929	93,854,070	2,533
116	209,478,351	376,294,029	8,105,970	9,808

Table 7 shows the distance traveled every 10 hours. The "Distance Traveled" column shows the total distance traveled. All distances are in meters. The "Remaining Distance " column shows the distance remaining before impact. The "Velocity" column shows the velocity of the moon at the completion of the specified time frame. We can see that the velocity of the moon continues to increase until it comes to a sudden halt on impact. The total time of travel in its fall was just over 116 hours. That would be 4 days and 20 hours of lapsed time before impact. At the time of impact the moon would be falling at a speed greater than 9.8 kilometers per second. That would be 21,941 miles per hour.

  Figure 5 is a chart of the distance traveled in the time frame of the fall.
We can see that the distance traveled increases dramatically over time.
Also notice, although units of measurement may be different, the graphical
curve describing moon fall is essentially the same as the one for free fall.
The big difference between the graphs is that the acceleration of the moon
is increasing over time, whereas Figure 4 – Free Fall, is over a short range
so the acceleration is constant.

Figure 5 – Distance Traveled

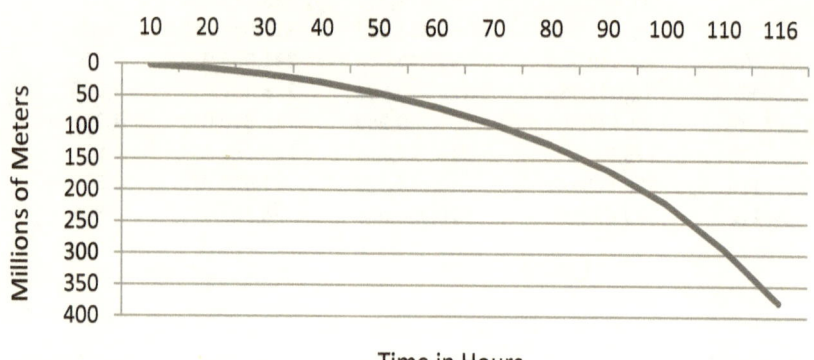

Time in Hours

Figure 6 shows the relative position of the moon after each 10 hour time period.

Figure 6 – Moon Fall

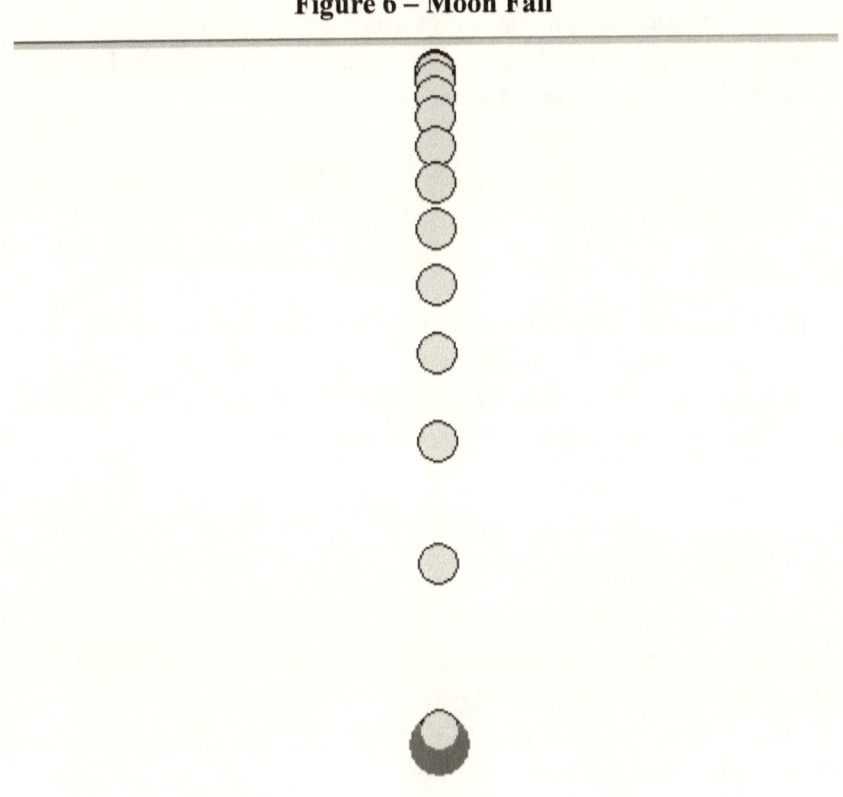

The relative sizes of the Earth and moon have been increased and are not shown accurately to scale. However, the relative position of each at the 10 hour interval is depicted accurately. This illustrates the moon's increasing speed and distance traversed within each time frame.

Space Flow

We have just examined the motion of bodies in free fall. This is all caused by gravity; and gravity is the effect of contracting space. Remember, gravity is not a force, it is an effect. Now let's examine the flow of space as it contracts toward some massive body. Let's take a look at the flow of space from the moon toward the Earth.

Before discussing the details of space flow let's remember the units of measurement of the gravitational constant.

$$G = 6.6742 \times 10^{-11} \quad \frac{\text{meters}^3}{\text{kilogram} * \text{second}^2}$$

We had previously stated we could interpret the Gravitational Constant to be a measurement of a specific volume of space being contracted by a kilogram of mass. If we multiply the Gravitational Constant by mass, then we can interpret GM to be a rate of space contraction.

$$GM = \frac{\text{meters}^3}{\text{second}^2}$$

Considering this, and that the Earth is absorbing space, we can now examine the flow of space from the moon to the Earth.

Table 8 contains the same kind of data that was shown in Table 7; however this data pertains to the flow of space itself, not an object in space. Space is flowing toward the Earth because of the contraction of space by the Earth. The speed at which space is flowing toward the Earth is shown in the last column and is given in millimeters per second. The distance in other columns is given in meters. The time interval is in millions of hours. From this table we can see that the flow of space is relatively slow compared to the bodies contained in the space. The bodies in space accelerate quite rapidly, while space flows slowly until it gets very close to a massive body.

The values for the speed of space flow in the last column of Table 8 can be calculated by the gravitational acceleration equation using the mass of the Earth and the distances in the "Remaining Distance" column.

$$a = \frac{Gm}{d^2}$$

Table 8 – Space Flow

Hours	Distance per time frame	Accumulated Distance	Remaining Distance	Speed in mm /sec
0	0	0	3.8440E+08	2.698
1,000,000	9.9674E+06	9.9674E+06	3.7443E+08	2.844
2,000,000	1.0528E+07	2.0496E+07	3.6390E+08	3.011
3,000,000	2.1143E+07	3.1671E+07	3.5273E+08	3.205
4,000,000	2.2461E+07	4.3604E+07	3.4080E+08	3.433
5,000,000	3.3975E+07	5.6436E+07	3.2796E+08	3.707
6,000,000	3.6385E+07	7.0360E+07	3.1404E+08	4.043
7,000,000	4.9257E+07	8.5642E+07	2.9876E+08	4.467
8,000,000	5.3414E+07	1.0267E+08	2.8173E+08	5.023
9,000,000	6.8639E+07	1.2205E+08	2.6235E+08	5.793
10,000,000	7.6182E+07	1.4482E+08	2.3958E+08	6.946
11,000,000	9.6825E+07	1.7301E+08	2.1139E+08	8.922
12,000,000	1.1498E+08	2.1180E+08	1.7260E+08	13.384
13,000,000	1.7519E+08	2.9017E+08	9.4234E+07	44.900
13,194,328	2.0284E+08	3.7803E+08	6.3710E+06	9,823.115

The rate of flowing space at any given location in space is equal to the acceleration of a body in space in that location. Let's take a look at the distance of the moon from the Earth – 3844x10[8] meters. At that distance the moon is accelerating toward the Earth at 2.698 millimeters per second squared. Space at that distance is flowing toward the Earth at 2.698 millimeters per second. Objects accelerate at a rate per second squared. Space flows at a steady rate per second. Objects accelerate every second, space flow does not. But, it's the steady flow of space that gives acceleration to the bodies contained in that space. So the last column of Table 8 could be interpreted as the acceleration applied to objects in that space, and it's also the speed of the flowing space.

Let's compare this to a jet airplane taking off from a standing position. The jet engines are turned up to full thrust and the breaks are released. At first the airliner barely begins to move. The thrust of the jet engines remains exactly the same; but second after second the airliner picks up speed accelerating down the runway. Near the end of the runway the airliner is capable of taking off. It lifts it's nose, the wheels come off the ground and it ascends into the sky. This all happens in a short period of time – from stand still to flight.

The airplane's speed increases every second because the force is applied continually every second. Every day multiple thousands of people stake their lives on this happening. So it is with space and the objects in space. Relate the flow of space to the thrust of the jet engines and the objects in space to the jetliner. It's the steady consistent flow of space that gives acceleration to the objects in space.

Figure 7 shows the flow of space from the moon to the Earth. It takes over 13,194,328 hours (1,505 years), for the space at the moon to reach the surface of the Earth. Yet with the continuous acceleration of the objects in space, it would take only 4 days and 20 hours for the moon to reach the Earth. The chart below shows the distance traveled by space in millions of hours. One million hours is about 114 years.

Figure 7 – Accumulated Distance

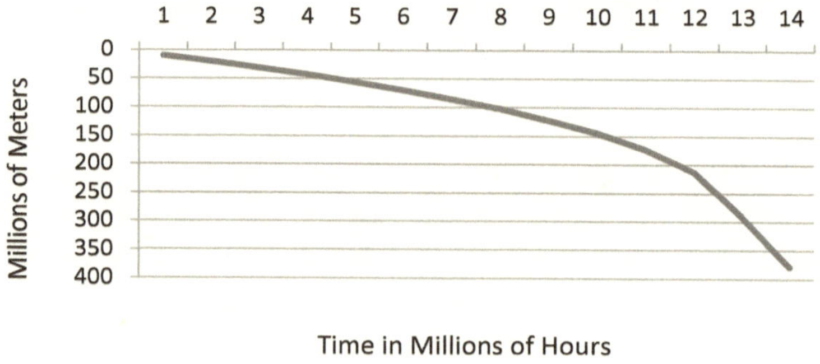

Time in Millions of Hours

Figure 8 shows the speed of space flow over time. For the first 11 million hours (about 1255 years), space is flowing below 10 millimeters per second. It obviously takes a long time to traverse large distances at such a slow rate of speed. The rate of space flow increases as it gets closer to the source of attraction. By the time space reaches the surface of the Earth, it's rate of speed is 9.823 meters per second. This is also the acceleration rate at ground level on the Earth. Space also flows according to the inverse square of distance law.

Figure 8 – The Rate of Space Flow

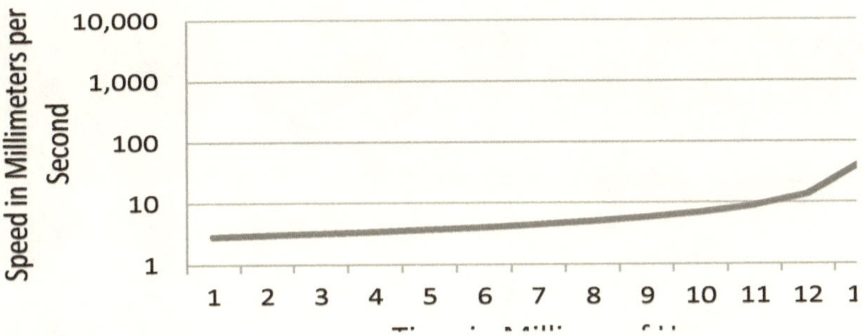

Space Flow Summary

What has been described in this chapter is the flow of space, and not only the flow of space, but the effect that it has on the objects in space. We've compared the flow of space to free falling objects in space. We've seen the flow of space to be relatively slow and steady, always toward objects of mass. Mass attracts space. We've referred to that phenomenon as the contraction of space, it is commonly called gravity. Space is contracting toward mass; we can also say that space flows toward mass. We've seen that the consistent flow of space causes an acceleration in the objects in space. Objects in space accelerate in the direction of the flow of space; they are not carried by the flow of space. It is the contraction of space that gives acceleration to the objects in space.

We've seen that the calculated rate of acceleration of objects in space is also the rate of speed at which space is flowing. Every second the rate of contracting space is added to the velocity of objects in space, giving them acceleration. This is a consistent phenomenon which works the same throughout the entire universe. It works all the time, every time, without fail, and cannot be blocked by any means. The effect of this phenomenon is what we call gravity.

Chapter 7

The Bending of Light

Light curves with the space it flows through.

The Curvature of Light

Einstein knew that somehow space was curved by massive objects. He saw light and all electromagnetic waves as energy, and energy as another form of matter. If that was true then light waves would follow the curvature of space just as matter does. Einstein predicted that a beam of light passing nearby the Sun would bend about 1.75 arc-seconds of curvature toward the Sun.

In 1919, two teams of astronomers took measurements to test Einstein's prediction during a total eclipse of the Sun. Their measurements, though inconclusive, were close enough to give credibility to the theory of relativity. Current technology makes more accurate measurements possible and confirms Einstein's prediction.

From these measurements we see that energy (in the form of electromagnetic waves) and matter change their direction according to the curvature of space. This was not predicted prior to Einstein because the assumption was that gravity was a force of attraction between objects of mass. So there was no thought of gravity having an effect on light. It should be noted that Bavarian astronomer Henry Cavendish and German physicist and astronomer Johann Georg von Soldner had independently calculated similar results; only they both calculated about one-half of the value predicted by Einstein. Cavendish's calculations (developed around the year 1784) went unpublished. Soldner's were published in 1803.[10]

These measurements show that gravity is not a force acting on objects. As space contracts toward massive objects, nearby objects and light waves turn toward the direction of the space flow. We can calculate the approximate curvature of a light wave passing near the Sun by calculating

The Fundamental Force

the amount of space contraction that takes place as the light passes. This would show that light is actually following the flow of space.

The diameter of the Sun is 13.92×10^9 meters. The speed of light is 2.9979×10^8 meters per second. Divide the diameter of the Sun by the speed of light and we get the number of seconds it takes a photon to pass by the Sun. That would be 4.6432 seconds. The table below shows the amount of curvature of a beam of light passing by the Sun at five one second intervals.

Table 9 – Curvature of Light

Second	distance	acceleration	cosine	curvature
1	9.1865E+08	157.30	0.7576	119.18
2	7.5782E+08	231.15	0.9184	212.30
3	6.9600E+08	274.04	1.0000	274.04
4	7.5782E+08	231.15	0.9184	212.30
5	9.1865E+08	157.30	0.7576	119.18
				936.99

The "distance" column is the distance that a ray of light is from the center of the Sun. The "acceleration" column shows the acceleration of space toward the Sun at the specified distance. The "cosine" column is the ratio of the radius of the Sun to the distance of the light ray. The curvature of the light ray is equal to the acceleration times the cosine. This gives us the amount of curvature in the direction of the Sun.

The total amount of curvature during the 5 seconds is 937 meters. We'll divide that by the radius of the Sun, 6.96×10^8 meters, to get the ratio of change in curvature – 1.346×10^{-6}. Multiple that by 360, 60, and 60 again to get the number of arc-seconds of curvature. The result is 1.745 arc-seconds of curvature; a close approximation to the 1.75 arc-seconds of Einstein's prediction. This simply illustrates that the curvature of light near massive objects is the result of the contraction of space. Electromagnetic waves and objects travel in a straight line. But that straight line is curved by space itself – as space contracts. That is what Einstein saw.

Although the paths of electromagnetic waves and objects of mass are affected by the curvature of space, there is a very distinct difference in the paths taken. Objects of mass will accelerate toward the source of contraction, thus gaining velocity and momentum. Electromagnetic waves will follow the path of the contracting space. Electromagnetic waves will not increase their velocity as they have neither mass or momentum.

Schwarzschild Radius

Until now we have been examining the flow of space as it approaches the Earth from the moon. Comparatively, this is a small scale flowing at a very slow pace. On a grander scale, when dealing with extremely massive objects, the rate of space flow increases with proximity to those objects.

Shortly before he died in 1916, Karl Schwarzschild (1873-1916) a Jewish German physicist and astronomer, solved Einstein's field equations of general relativity. Einstein gave us an approximate solution to his field equations, but it was Schwarzschild who used polar coordinates to produce an exact solution. But Schwarzschild discovered much more from the field equations. He realized that Einstein's equations made possible the existence of something later referred to as a "Black Hole." The name aptly described an object of such enormous mass that it swallows everything that comes near – including light. There has been much debate among physicists whether black holes can actually exist in the real world. At first even Einstein did not think that possible.

Schwarzschild calculated the necessary radius of an object from which nothing would escape. This radius is now referred to as the Schwarzschild Radius, and is sometimes used to calculate the event horizon. Schwarzschild used Equation 10 – Escape Velocity to aid in the determination of the Schwarzschild Radius. The escape velocity equation gives us the minimum velocity required to escape the gravitational attraction of a body to which we are surface bound. That equation is shown below:

Equation 10 – Escape Velocity

$$v = \sqrt{\frac{2Gm}{r}}$$

Where:

    v   is the escape velocity
    G   is the Gravitational Constant
    m   is the mass of the object from which to escape
    r   is the radius of that object

Schwarzschild used Equation 10 to calculate the radius of a massive object when the escape velocity is set to the maximum speed limit - the speed of light. He then solved the equation for 'r' to get the radius at which there is no escape. One cannot exceed this speed so there is no escape, not even for light. If even light cannot escape the gravitational clutches of a massive body, then this is the radius within which a black hole resides. The equation to calculate the Schwarzschild Radius is shown below:

Equation 11 – Schwarzschild's Radius

$$r_s = \frac{2Gm}{c^2}$$

Where:

    r_s is the Schwarzschild Radius
    G   is the Gravitational Constant
    m   is the mass of the object from which to escape
    c   is the speed of light

The Event Horizon

Closely related to Equation 11 – Schwarzschild's Radius is another equation used to calculate a critical radius referred to as the event horizon. The term "event horizon" refers to the location in space surrounding a black hole from which light cannot escape. Here is where the speed at which light traveling outward is the same at which space is approaching the source of the light. When the flow of space into a massive object equals the speed of light radiating from the object, light cannot escape from within that radius. At this point, light from within the bounds of the event horizon would appear to be standing still, as if frozen in time and space.

The equation to determine the event horizon was derived from the acceleration equation (Equation 6) we have been using since the Gravity chapter. It was derived in the same manner as Schwarzschild's Radius. The notion for the event horizon equation is that the radius at which nothing could escape the grip of gravity is the radius at which the gravitational attraction is equal to the speed of light.

The equation below is the acceleration equation only with the acceleration set to the speed of light (c):

$$c = \frac{Gm}{r^2}$$

Solving this equation for 'r' gives us the critical radius at which the speed of space flow is equal to the speed of light. The radius at which this occurs is the event horizon and can be calculated using the following equation:

Equation 12 – Event Horizon

$$r_e = \sqrt{\frac{Gm}{c}}$$

Where:
  r_e is the radius of the event horizon
  G   is the Gravitational Constant
  m   is the mass of the object from which to escape
  c   is the speed of light

Figure 9 shows the Schwarzschild Radius in gray (calculated from Equation 11), and the event horizon in black which was calculated from Equation 12 – Event Horizon.

The Fundamental Force

Figure 9 – Schwarzschild Radius & Event Horizon

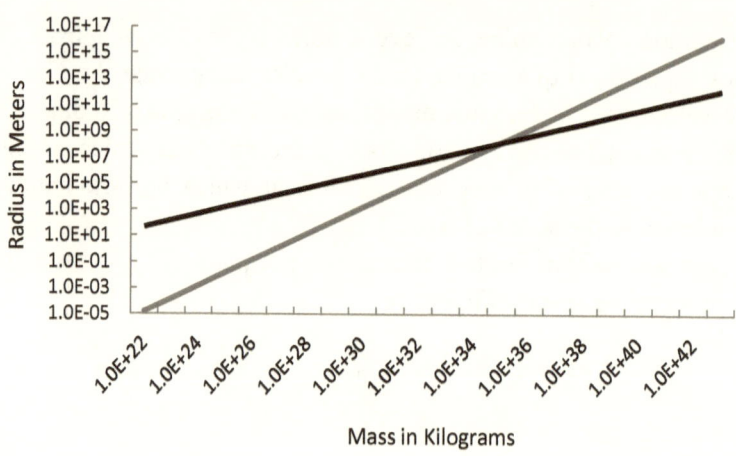

Figure 9 shows that the Schwarzschild Radius of the attracting mass increases proportionately with mass. For every increase of magnitude of mass there is a corresponding increase in the magnitude of the Schwarzschild Radius. This shows that the Schwarzschild Radius is not related to the Law of Inverse Squares. The event horizon of the attracting mass (in black) increases in magnitude with the square root of the mass's increase of magnitude and therefore abides by the Law of Inverse Squares.

Validating Assumptions

One might have expected the lines on this graph to be the same. One of them is not valid. Because of the differences in the graph we need to compare and validate the radius produced by each equation. We'll use the mass of our own Sun for this test case. The mass of the Sun is 1.989×10^{30} kg. Multiplying the Sun's mass by the gravitational constant is equal to 1.3275×10^{20}; so the value of Gm is 1.3275×10^{20} m^3/s^2. The calculation for the radius produced by each equation is shown below:

Schwarzschild Radius

$$r_s = \frac{2\,Gm}{c^2}$$

$$2{,}954 = \frac{2 * 1.3275\text{x}10^{20}}{c^2}$$

Event Horizon

$$r_e = \sqrt{\frac{Gm}{c}}$$

$$665{,}437 = \sqrt{\frac{1.3275\text{x}10^{20}}{c}}$$

The calculation for the Schwarzschild Radius of the Sun is 2,954 meters. If all the mass of the Sun were compacted into a sphere with a radius of 2,954 meters it would be a black hole; nothing would be able to escape from its surface. But if the radius were a little larger like 3,000 meters, then light should be able to escape.

The Schwarzschild Radius is based on the escape velocity equation, with no consideration of space flow. It's interpretation may not be accurate. Let's see if the Schwarzschild Radius holds up to critical scrutiny. Let's use the acceleration equation to validate the Schwarzschild Radius:

$$a = \frac{Gm}{r^2} = \frac{6.6742\text{x}10\text{x}^{-11} * 1.989\text{x}10^{30}}{(3{,}000)^2} = 1.475\text{x}10^{13}\ m/s^2$$

The acceleration produced by the Schwarzschild Radius equation is much greater than the speed of light. So even at 3000 meters there is no escape!

The calculation for the event horizon using Equation 12, gives a radius of 665,437 meters. Let's try that radius with the acceleration equation:

$$a = \frac{Gm}{r^2} = \frac{6.6742\text{x}10\text{x}^{-11} * 1.989\text{x}10^{30}}{(665{,}437)^2} = 2.9979\text{x}10^8\ m/s^2$$

Yes, that gives us an acceleration equal to the speed of light. Which means that if all the mass of the Sun were compacted into a sphere with a radius of 665,437 meters, nothing on the surface could escape from this hypothetical black hole. However beyond a radius of 665,437 meters light would be able to escape. Therefore, Equation 12 – Event Horizon is the accurate method to calculate the event horizon.

One other consideration must be taken into account. Examine Figure 9. The two radii match at about a mass of $1.0x10^{35}$ kilograms; beyond that the Schwarzschild Radius is larger than the event horizon. Let's do the same calculations again only using a mass of $1.0x10^{43}$ kilograms; that would be 50 billion times more massive than our Sun. Using Equation 11 to calculate the Schwarzschild Radius would give a radius of $1.4852x10^{16}$ meters. The calculation for acceleration at the Schwarzschild Radius follows below:

$$a = \frac{Gm}{r^2} = \frac{6.6742x10x^{-11} * 1.0x10^{43}}{(1.4852x10^{16})^2} = 3.0257 \text{ m/s}^2$$

The acceleration produced by Schwarzschild's Radius is only 3.0257 m/s^2, which is much lower than the speed of light. Not only that, but it's three times less than the gravitational attraction we feel on the surface of the earth. Which means there is no black hole here, and the Schwarzschild Radius cannot be valid for masses over $1.0x10^{35}$ kilograms.

The calculation for the event horizon using Equation 12 gives a radius of $1.4921x10^{12}$ meters. This is 10,000 times smaller than Schwarzschild's Radius. Now let's use the event horizon radius with the acceleration equation:

$$a = \frac{Gm}{r^2} = \frac{6.6742x10x^{-11} * 1.0x10^{43}}{(1.4921x10^{12})^2} = 2.9979x10^8 \text{ m/s}^2$$

Yes, once again that gives us an acceleration equal to the speed of light. Which means that if all $1.0x10^{43}$ kilograms were compacted into a sphere with a radius smaller than $1.4921x10^{12}$ meters, nothing could escape its gravitational attraction, not even light. Equation 12 gives us an accurate radius for the event horizon.

Traveler's Advisory

The difference between the Schwarzschild Radius and the event horizon is the equations they are based on. Schwarzschild's Radius is based on the escape velocity from the surface of the attracting mass. That may not be a valid assumption; it does not take into account the acceleration of space or the Law of Inverse Squares. The event horizon equation is based on the law of gravity. Any mass which accelerates the flow of space to the speed of light is a black hole.

This is a warning of hazardous conditions to all future space travelers who would be tempted to venture near black holes. Do not use the Schwarzschild Radius as the event horizon! Use Equation 12 – Event Horizon, for your calculations. Also take along a copy of Figure 9 – Schwarzschild Radius & Event Horizon. Do not leave home without these, and do not cross the black line.

We can now see that all electromagnetic waves follow the flow of space. Light waves are bent with the flow of space, and can be bent drastically when passing massively large objects.

Chapter 8

Space Consumption

Matter has an insatiable appetite for space.

The Consumption of Space

In the previous chapters on gravity we left some questions unanswered. What happens to all that absorbed space? Just as a sponge soaking water reaches a saturation point, does matter have a limit? Or will it continue to absorb space forever?

To answer these questions we must make another assumption. This new assumption is that matter also consumes space. So the new word is "consumption" and the new term is "space consumption." In this case we do consider space to be reduced in size and ultimately shrunken out of existence. And yes, this has been an ongoing process from the beginning. Mass consumes space! That's what it does by its very nature. There is no law against the consumption of space. Space is being created all the time. That's how the universe expands. With every passing moment the boundary of the universe increases, filled with new space. In each of those moments, a relatively small portion of space is being consumed by matter. This will, of course, require explanation. We will see shortly where the space is consumed.

Gravitational Flux

In the previous chapter Space Contractions, in the section, The Absorption of Space, we calculated the amount of space absorbed by the Earth. The answer we obtained after that long and laborious calculation was 5.0104×10^{15} cubic meters. In the section, The Law of Volumes, also in the Space Contractions chapter, we saw another equation that could be used to calculate the volume of space being contracted. All we needed to know was the radius of the objects orbit, and its acceleration. That equation is shown below.

$$v = 4\pi r^2 a$$

The equation for acceleration is:

$$a = \frac{Gm}{d^2}$$

If we substitute for acceleration (a) in the volume equation we get the following equation:

$$v = \frac{4\pi r^2 Gm}{d^2}$$

Since the radius (r) and the distance (d) are referring to the same thing, they cancel each other out. So we are left with Equation 13 in its final form. This is the equation for the volume of space contracted and absorbed by matter, and then consumed by matter.

Equation 13 – Volume Consumption

$$V_c = 4\pi Gm$$

Where:

- V_c is the volume of space contracted.
- G   is the gravitational constant.
- m   is the mass of the object contracting space.

This expression $4\pi Gm$ refers to something called "Gravitational Flux". So let's calculate the gravitational flux of the Earth, which is also the volume of space consumed by the Earth each second.

$$V_c = 4\pi * 6.6742 \times 10^{-11} * 5.974 \times 10^{24}$$

$$V_c = 5.0104 \times 10^{15} \text{ cubic meters per second squared}$$

This is the amount of space absorbed, and consumed, by the Earth every second. The quantity of absorbed space we previously calculated and the gravitational flux just calculated are identical. This is a most interesting connection and quite fortuitous. This means that we can easily calculate the space consumed by whatever massive object we're interested in. All we need to know is the mass of the object of interest. Now we can make some interesting calculations.

In the chapter on the Flow of Space we saw that the time required for the space at the moon's orbit to flow to the earth was approximately 1505 years. The cause of the space flow is the consumption of space by mass. We have just calculated the volume of space consumed by the earth every second. Now let's calculate the volume of space that occupies the sphere surrounding the Earth with a radius equal to the moon's orbit around the Earth:

$$ v = \frac{4\pi r^3}{3} = \frac{4\pi(3.844 \times 10^8)^3}{3} = 2.37924 \times 10^{26} \ m^3 $$

That volume is 2.37924×10^{26} meters cubed. Now let's divide that by the volume of space consumed by the earth every second to get the number of seconds needed for the Earth to consume all the space out to the moon.

$$ \frac{2.37924 \times 10^{26}}{5.0104 \times 10^{15}} = 4.7486 \times 10^{10} \ seconds $$

Finally we'll divide that by the number of seconds in a year to get the time of space flow from the moon to the Earth in years:

$$ \frac{4.7486 \times 10^{10}}{3.15576 \times 10^7} = 1504.7 \ years $$

The results of our calculations match up with what was presented in The Flow of Space chapter.

Until now we've been working with massive objects in space. Let's turn our attention to the basic particles of matter. The following table shows the mass and classical radius of each of the fundamental particles:

Table 10 – Fundamental Particles

	Mass	Radius
Electron	9.1094E-31	2.8179E-15
Proton	1.6726E-27	1.5347E-18
Neutron	1.6749E-27	1.5325E-18

Using Equation 13 let's calculate the volume of space consumed by the fundamental particles of matter:

Electron: $v_c = 4\pi * 6.6742 \times 10^{-11} * 9.1094 \times 10^{-31} = 7.6401 \times 10^{-40}$

Proton: $v_c = 4\pi * 6.6742 \times 10^{-11} * 1.6726 \times 10^{-27} = 1.4028 \times 10^{-36}$

Neutron: $v_c = 4\pi * 6.6742 \times 10^{-11} * 1,6749 \times 10^{-27} = 1.4048 \times 10^{-36}$

We have just penetrated the barrier into the subatomic world. We have just calculated the space consumed each second by the basic particles of matter. Right now this of and by itself does not give us very much information. But remember, it's just one piece at a time.

Consumption Rate

Now we can resume with some interesting and revealing calculations. Let's take a look at the ratio of volume of space consumed to the mass doing the consuming. Let's divide the volume of space consumed by the Earth by the mass of the Earth.

$$\frac{\text{volume consumed}}{\text{mass of Earth}} = \frac{5.0104 \times 10^{15}}{5.974 \times 10^{24}} = 8.387 \times 10^{-10}$$

This is the relationship between the rate of space consumption and the mass doing the consuming. Doing this same calculation for the fundamental particles of matter gives the same results:

Electron: $7.6401 \times 10^{-40} \div 9.1094 \times 10^{-31} = 8.387 \times 10^{-10}$
Proton: $1.4028 \times 10^{-36} \div 1.6726 \times 10^{-27} = 8.387 \times 10^{-10}$
Neutron: $1.4048 \times 10^{-36} \div 1,6749 \times 10^{-27} = 8.387 \times 10^{-10}$

This shows that the fundamental particles of matter all consume space at the same rate. The ratio of volume of space consumed to mass is always the same. Let's call it the Space Consumption Rate. It is equal to:

Equation 14 – Space Consumption Rate

$$C_r = 4\pi G = 8.387 \times 10^{-10} \ \frac{m^3}{kg \ s^2}$$

Where:
C_r is the consumption rate of space.
G is the Gravitational Constant.

The expression $4\pi G$ is equal to 8.387×10^{-10} cubic meters per kilogram per second squared. This is the rate of space consumption. One kilogram of mass, (about 2.2 pounds), would consume just 8.387×10^{-10} cubic meters of space each second. That is a little difficult to visualize. We can make this relationship a little easier to comprehend by multiplying the volume of space consumed by one billion (1000^3). This may give us a better perspective because this will show the amount of space consumed in cubic *millimeters*. One kilogram of mass consumes just .8387 cubic *millimeters* of space each second. Also one pound of mass consumes about .38 cubic *millimeters* of space every second. That's not very much space being consumed, it fact it would seem insignificant. But the effect of that space consumption is gravity.

Energy Density

If we took the mass of the Earth and multiplied it by the speed of light squared, we would have the sum total of the Earth's energy. Remember Einstein's equation: $E = mc^2$. If we divided the energy contained within the Earth by the volume of space consumed, we would then have the energy density of the Earth. Let's do the calculation for the Earth:

$$\frac{\text{energy of Earth}}{\text{volume consumed}} = \frac{5.974 \times 10^{24} * c^2}{5.0104 \times 10^{15}} = 1.0716 \times 10^{26}$$

If we took the energy of each fundamental particle and divided it by the volume consumed, we would have the energy density of that particle.

Electron: $9.1094 \times 10^{-31} * c^2 \div 7.6401 \times 10^{-40} = 1.0716 \times 10^{26}$
Proton: $1.6726 \times 10^{-27} * c^2 \div 1.4028 \times 10^{-36} = 1.0716 \times 10^{26}$
Neutron: $1,6749 \times 10^{-27} * c^2 \div 1.4048 \times 10^{-36} = 1.0716 \times 10^{26}$

These calculations show that each of the fundamental particles has the same amount of energy per volume consumed. In fact, since all matter is made from these fundamental particles, the energy density of all matter is the same. Listed below is the equation for energy density.

$$E_d = \frac{m_x c^2}{4\pi G m_x} \qquad \frac{\text{energy in joules}}{\text{volume consumed}}$$

The mass cancels out of the equation so we are left with the equation (Equation 15) in its final form:

Equation 15 – Energy Density

$$E_d = \frac{c^2}{4\pi G} \qquad \frac{\text{speed of light squared}}{\text{space consumption rate}} \qquad \frac{\text{Kg}}{\text{m}}$$

Where:

    E_d is the energy density.

    G   is the Gravitational Constant.

    c   is the speed of light.

The Source of Consumption

We have seen how the contraction of space is the cause of the attraction of all matter toward all matter. Gigantic bodies in space are attracted to each other by what has been called gravity, or the gravitational force. Yet, now we can see it's not a force at all. It's the effect of space consumption! Giant bodies in space move according to the flow of space. The flow of space results from the consumption of space by all matter. We have seen that space is absorbed into the Earth, but what happens to it? We have calculated the consumption rate of space for all the fundamental particles of matter, and found it to be a constant. Now, we will examine the source of the consumption. By "source" I mean the prime cause of the consumption.

In the Space Contractions chapter, we saw an equation we could use to calculate the volume of space being attracted to some object. We used that equation to calculate the volume of space being contracted toward and absorbed into the Earth. Now we will use that equation with just a small adjustment to calculate the space consumed by the fundamental particles of matter. In this case we will eliminate the 4π from the equation, this will give us the exact volume of space consumed. The equation is shown below:

$$V_c = r^2 a$$

First we need to calculate the acceleration toward those particles using the basic acceleration equation. The first calculation is for an electron. The mass of an electron is 9.1094×10^{-31} kilograms, and the classical radius of an electron is 2.81794×10^{-15} meters.

$$a = \frac{GM_e}{d^2} = \frac{6.6742 \times 10^{-11} * 9.10938 \times 10^{-31}}{(1.46885 \times 10^{-13})^2} = 2.81794 \times 10^{-15}$$

The second calculation is for a proton. The mass of a proton is 1.6726×10^{-27} kilograms, and the classical radius of a proton is 1.5347×10^{-18} meters.

$$a = \frac{GM_p}{d^2} = \frac{6.6742 \times 10^{-11} * 1.67262 \times 10^{-27}}{(2.69703 \times 10^{-10})^2} = 1.53470 \times 10^{-18}$$

Notice that the accelerations of the two most fundamental particles, at the designated distances, is the same as their classical radius. Now we can use the space consumption equation to calculate the exact amount of space each particle consumes. Once again the first calculation is for an electron:

$$v_c = r^2 a = (1.46885 \times 10^{-13})^2 * 2.81794 \times 10^{-15} = 6.07978 \times 10^{-41}$$

Next is the calculation for a proton:

$$v_c = r^2 a = (2.69703 \times 10^{-10})^2 * 1.53470 \times 10^{-18} = 1.11633 \times 10^{-37}$$

Now notice that results of these last two calculations is exactly the same as the product of Gm. Those calculations are repeated below:

$$Gm = r^2 a$$

Electron: $Gm = 6.6742 \times 10^{-11} * 9.10938 \times 10^{-31} = 6.07978 \times 10^{-41}$
Proton: $Gm = 6.6742 \times 10^{-11} * 1.67262 \times 10^{-27} = 1.11633 \times 10^{-37}$

We see that the volume of space being consumed can be calculated in two different ways. The advantage of using ($v_c = r^2 a$) is that the r^2 tells us at what distance from center the acceleration is occurring. For an electron consuming space the acceleration is 2.81794×10^{-15} and its distance from center is 1.46885×10^{-13}. The amount of space consumed is 6.07978×10^{-41} Cubic meters.

The surface area of a sphere can be calculated using the expression $4\pi r^2$. Envision the space being consumed as a hollow sphere like a balloon.

The surface of the balloon is at a radius of 1.46885×10^{-13} meters. The thickness of the balloon's surface is it's acceleration – 2.81794×10^{-15} meters. The expression ($4\pi r^2 a$) gives the volume of the space surrounding the electron. The volume for the electron is 7.6401×10^{-40} cubic meters per second squared. Divide that by 4π and that gives the amount of space consumed by the electron! The space is consumed out of existence. This is a continuous process that does not stop. The appetite of matter for space is in fact insatiable.

In this chapter we have come to the end of the trail, for right now, concerning the contraction of space. But, there is more to come later. Space is contracting toward matter, then absorbed by the matter, then consumed by the matter. Looking at it another way we could say, matter is consuming space, causing more space to be absorbed, causing space to be contracted toward the matter, and ultimately causing objects to be attracted to each other. And that is the cause of gravity! This consumption must go on forever, otherwise gravity would cease to be.

Chapter 9

Dark Energy

Dark Energy is there – we just can't see it.

The Cosmological Constant

When Einstein developed his General Theory of Relativity, he also developed equations that would approximately model how the universe worked. These equations are referred to as Einstein's Field Equations. They contain things like Euclidean Geometry, Lorentz Transformations, and Riemann Metric Tensors. Although very complex, these mathematical expressions are necessary to account for the curvature of space, to make vector arithmetic work correctly; and for whatever other reasons they like. It's all okay with me. It all makes Newton's equations look so simple and so sweet. We do not need to use Einstein's Field Equations here, but it's good to know they exist, and that there are people who understand them.

One day Einstein was mulling over in his mind the implications of his equations when a bad thought occurred. He remembered a problem Newton had discovered many years before, and which was verified by French mathematician Simeon-Denis Poisson. Newton realized that his discovered force of gravity would eventually over time attract all matter together at the center of the universe. If there is no repelling force to keep matter spread out throughout the universe, then gravity would eventually attract all matter together. This would be true no matter how widespread the matter. All matter from all over the universe would ultimately come together in one place – often referred to as "The Big Crunch."

This was a bit of a dilemma for Einstein. At the time he thought the universe was static (often called stationary), not expanding or contracting. Remember at that time there was no concept of anything existing outside the Milky Way. The universe was the Milky Way and all it contained; and the Milky Way seemed to be in a steady state. Einstein realized he needed to change the gravitational field equations to include a constant that would prevent the universe from collapsing in on itself. This constant was referred to as the cosmological constant. If the cosmological constant had a value of zero, then the field equations would work as originally defined. A negative value would depict a collapsing universe. A positive value implied a force that counteracts gravity and would depict an expanding universe. Einstein cleverly protected his field equations with the cosmological constant. Thus the equations would be valid no matter what the universe did.

Just after making this ingenious little change a friend of his, Dutch cosmologist Willem De Sitter, came up with a solution to Einstein's equations - without the cosmological constant. But in De Sitter's solution there appeared to be a rather peculiar phenomenon causing the universe to expand. At first Einstein could not accept the truth contained in De Sitter's solution. He thought De Sitter had made an error; but he could not disprove De Sitter's solution or conclusion.

Perhaps, Russian mathematician Alexander Friedmann (1888-1925) helped Einstein reconsider his position. Einstein based the development of his field equations of gravitation on two hypotheses:[11]

1. There is an average density of matter throughout the whole universe and the average density is the same everywhere.
2. The size of space, (the universe), is independent of time.

Friedmann suggested a solution to the equations if the second hypotheses were modified or dropped. The second hypotheses stated that the magnitude of space, (the size of the universe), is independent of time. The solution that Friedmann proposed did not require the cosmological constant; but it did require that the magnitude of space depends on time (i.e. expanding space). In other words, the universe had to be expanding with time. According to Alexander Friedmann, the theory of relativity demands an expansion of space, and an expanding universe.[12]

Now after much thought and deliberation, Einstein was convinced. He realized that the universe must be expanding. His field equations were correct. The solution that Willem De Sitter presented was correct. Alexander Friedmann's analysis was also correct. After re-evaluating the solutions presented, Einstein concluded that the cosmological constant was nothing but a big blunder.

Previously, Einstein had strongly believed that the universe was static, not expanding or contracting. Now, based on his mathematical equations, the geometry of space and time, and his own insight, and a little help from his friends, he changed his position about the state of the universe. He came to realize that his field equations implied an expanding universe. There was no proof – yet. But the proof would come shortly.

The Standard Candle

In the beginning of the 20th century it was believed that the entire universe was contained within the Milky Way. Everything that was seen, either with a telescope or the naked eye, was seen as part of the Milky Way. There was no accurate way to gage the distances of stars.

American astronomer Henrietta Leavitt (1868-1921), changed all that. She made a significant breakthrough by discovering a way to accurately calibrate the distances of a certain kind of star. Her discovery began by studying stars that would vary in brightness. Over a regular period of time these stars would cycle in their degree of brightness. These kinds of stars were naturally called variables. The first stars she discovered having these peculiar characteristics where in the constellation Cepheus, so they became known as Cepheid variables.

Leavitt was able to accurately calculate the distance of the Cepheid variables based on their period and brightness. This type of star became known as "the standard candle". Once the brightness of a candle is known, then its distance can be determined by the amount of light received from it. This is how Cepheid variables can be used in determining the distances of far way galaxies.

Hubble to the Rescue

Edwin Hubble (1889-1953) was a brilliant American astronomer, who changed the way we see the universe. Hubble started analyzing the light from Cepheid variables. The reason for selecting Cepheid variables was because their distances from the observer could be determined with a high degree of accuracy. By analyzing their distances, he could get a better idea of the size and shape of the Milky Way. Hubble began his investigation to learn more about the size and scope of the Milky Way, but what he found was something entirely unexpected.

One of the Cepheid variables under study was located in the Andromeda Nebula. At that time Andromeda was considered a nebula within the Milky Way. Analysis on his star showed it to be far outside of the Milky Way galaxy. This was a startling discovery that astounded the scientific world. Andromeda was not a nebula; it was another galaxy, far outside the Milky Way.

This discovery showed that there was much more to the universe than just the Milky Way. Hubble's survey of the Milky Way then led to the discovery of many more galaxies. They were just beginning to discover what we now know. The universe was much larger than anyone had previously imagined. The Milky Way was just one galaxy in a universe filled with billions and billions of galaxies. This discovery by Hubble did in fact totally change the way the universe was viewed.

Hubble wasn't finished; he discovered even something more interesting. He continued analyzing all the Cepheid variables he could find in other galaxies. By using them he was able to determine the distances of far away galaxies. Now he was able to classify galaxies by size, shape, and distance. In his analysis of the galaxies he also used a technique called the Doppler Effect. This technique enabled him to make another startling discovery.

The Doppler Effect is named after Austrian mathematician and physicist Johann Christian Andreas Doppler (1803-1852). Doppler identified and documented the perceived change in frequency and wavelength of waves caused by the relative motion of either the observer or the source of the wave under observation. This effect can be applied to most any kind of wave, especially sound waves and light waves. The Doppler Effect is used in police radar guns to detect the speed of vehicles. Hubble found a much better use of the effect - to calculate the direction and speed of galaxies. The light we perceive from galaxies would be changed according to their relative motion.

The Doppler Effect gave a way of determining whether a star was moving away from the observer, or toward the observer. If a star was moving away from the observer, the light from the star would be shifted toward the red side of the color spectrum. If the star was moving toward the observer, it's light would be shifted toward the blue side of the spectrum. Also, the degree of shift would be a good indication of the speed in which the star was traveling through space. The greater the shift toward the red side of the color spectrum, the faster the star was moving away from the observer. The greater the shift toward the blue, the faster it was moving toward the observer.

Star after star showed its color to be red shifted. This meant that the galaxies under study were moving away from us. He continued analyzing more stars in every direction, usually with the same results – red shifted. Only galaxies in a local group could be moving toward each other; but beyond the local group, with very few exceptions, they were red shifted.

The general trend Hubble observed was that the galaxies seemed to be drifting apart from each other. In fact there was a direct relationship between the rate of recession and the distance from us; the farther away the galaxy, the greater the redshift, which meant the faster the recession. Hubble plotted the trend of the galaxies he studied. The conclusion Hubble drew was clear – the universe was expanding at a constant rate. The value of the expansion rate is known as Hubble's constant. We will discuss Hubble and his findings in more detail in a later chapter.

Since then, there has been much speculation on the fate of the universe. Would the expansion slow down and eventually come to a halt; after which the force of gravity would bring all matter back together for another big bang? That's not a pretty picture. This seemed to be the prevalent view, perhaps because the alternative was not very attractive, or worse. The alternative would be that the universe would continue to expand forever with all matter scattered throughout the universe. Eventually all galaxies would recede out of sight. Eventually all stars would exhaust their fuel supply and turn into cold dead bodies. Eventually entropy wins and everything turns cold and lifeless. That's not a pretty picture either. However, that's not the end of the story. The fate of the universe will also be discussed in a later chapter.

The Type Ia Supernova

Saul Perlmutter of the University of California at Berkley lead a team of astronomers from the Lawrence Berkley National Laboratory in search of type Ia supernova. This project was referred to as the Supernova Cosmological Project (SCP). The search was on for supernova of type Ia. The Ia kind of supernova signifies the death of a white dwarf and its partner; a duel star system that explodes. There is a good reason to look for this kind of supernova. They make an accurate yardstick similar to the Cepheid variables that Hubble used in his analysis of the universe. This is the new "standard candle". The Type Ia supernova offer the advantage of being seen over much greater distances than the Cepheid variables. This allows for accurate measurements into the far reaches of the universe.

Using techniques that were similar to the techniques used by Hubble, but with new tools and new technology, Perlmutter's team was able to look far deeper into space than ever before. They were checking the redshift of Type Ia supernova far out in deep space. Perlmutter's intent was to measure the rate of deceleration of the universe – how fast was the expansion of the universe slowing down. When would it begin to contract?

When Einstein learned of Hubble's discoveries, it was confirmation that his field equations were correct. He no longer needed his cosmological constant. But now we do! Why? Because of Perlmutter's discovery – the universe is expanding at a faster rate. Now there is a big search on by cosmologists all over the world; a search to find this strange and powerful new force that's causing the expansion of the universe. This mysterious force has been hidden from our view and understanding until recently. It is now referred to as dark energy. The dark energy is here and the cosmological constant has made a comeback. This time it's back to stay. The cosmological constant tells us the energy density of a vacuum, and it's related to the rate of expanding space.

Characteristics of Dark Energy

Just what is dark energy and why can't we see it? For the same reason we can't see radio waves, or micro waves. Visible light waves are only a small part of the spectrum of electromagnetic waves that we can see. Since we can't see energy, we have to look at the effects it causes.

Let's talk about the properties of empty space. First of all empty space is not completely empty. Empty space may be void of mass, but it contains electromagnetic waves that cannot be seen. Space is very pliable; space can be expanded, and space can be contracted. In the chapters on gravity I attempted to explain gravity in terms of contracted space. It would seem reasonable to me that if space can be contracted, then space can also be expanded. We can also say, according to Alexander Friedmann, that the magnitude of space increases with time.

Scientists are speculating that space is expanding because of the dark energy contained therein. Some consider this dark energy to be the mysterious force that's causing the size of the universe to be forever increasing. It is behaving just like Einstein's cosmological constant. As the dark energy increases, so does the space containing it. So, the overall energy density of the universe remains the same.

It's the increasing magnitude of space that's causing the universe to grow larger. The universe is continually growing because the space contained therein is constantly expanding. Expanding space is the cause; an expanding universe is the effect.

So, now do we really have five known forces in the universe to contend with? This seems to be getting a bit too complicated. Perhaps all the fields of force can be simplified as we come to understand dark energy. I suggested in the chapter on gravity, that gravity was not a force but an effect. Perhaps the same applies to the other forces. What I am suggesting is that all four known forces in the universe are only effects of the one fundamental force that powers the whole universe. That fundamental force is currently going by the name "Dark Energy."

Chapter 10

The Pioneer Anomaly

It's just one anomaly after another.

The Pioneer Anomaly

In March of 1972 the Pioneer 10 space craft was launched. It's mission – to capture and send back to Earth information about the far reaches of our solar system. One year later, in April 1973, an identical space craft, Pioneer 11, was launched in the opposite direction. Its mission was the same as its sister, Pioneer 10. These two probes would feed us much valuable information about the planets and moons in our solar system.

With their mission successfully accomplished, their job was done. With fuel supply exhausted and battery power nearly discharged, they were ready to retire. These two spacecraft would now receive their reward. They would be left to soar freely out into deep space, far beyond the solar system at a cruising speed of about 27,000 miles per hour. In June 1983, over eleven years after its launch, Pioneer 10 became the first spacecraft to leave the solar system.[13] With no more planets to pass, no more pictures to take, there would be no more information to gather and nothing to report back to Earth. Their work was done and they were forgotten – or were they?

Not by John Anderson, an astronomer at the Jet Propulsion Laboratory in Pasadena, California. In the early 1980s Anderson tracked the whereabouts of these two tiny space probes as they traveled beyond the solar system. The Pioneer spacecraft (shown in Figure 10) are about 2.9 meters (9.5 feet) long, with a trim mass of 259 kg (571 pounds). Anderson noticed that both of these probes were slightly off their expected course. Both were beginning to slow down at a constant rate. Both probes were decelerating. The rate of deceleration is approximately 8.74×10^{-10} meters per second squared. This could also be thought of as an acceleration toward the Sun. Something was holding them back – but what?

Figure 10 – Pioneer 10-11

Image Credit: NASA

http://nssdc.gsfc.nasa.gov/image/spacecraft/pioneer10-11.jpg

The Search is On

Must be some kind of observational errors, was the first thought. Check measurements, check for computational errors, eliminate approximations, and then recalculate. Same results! Okay, recalibrate instruments, check the computer programs being used, check everything again and again, and then recalculate again. Same results!

The deceleration of the Pioneer spacecraft had become too big and complex of a problem for one person to solve. Anderson gets help from close friends, a theoretical physicist Michael Nieto from Los Alamos National Laboratory in New Mexico, and NASA scientist Slava Turyshev from the Jet Propulsion Laboratory in Pasadena. Now, the search is on for the explanation of this unexpected phenomenon. They repeat all the analysis and checks all over again. Same results! Check again? Yes – same results!

If this deceleration is real, then there must be a cause, and a plausible explanation. Was there a gas or a thermal radiation leak? Are there unidentified bodies of mass, perhaps in the Kuiper belt, causing the drag on the Pioneers? Could there possibly be undetected dark matter hiding out just beyond the solar system? This would be something like the halo of dark matter that some scientists suggest may be surrounding the spiral galaxies. What about the affects of solar wind and radiation, cosmic rays, or even space dust?

There are obviously many issues that would require detailed analysis to determine their effects on the two probes. More scientists from all over the world have joined in the effort to find the reason for the Pioneer anomaly. Meetings are held, information is gathered; and no stone is left unturned. All issues are under investigation, all are suspect, but none have yet been fingered as the culprit. To this day the problem remains and has sparked the interest of many others scientists and interested people. Some are willing to offer suggestions and opinions, but nothing conclusive has been forthcoming. All incoming information is collected and reviewed, every option is considered; but no convicting evidence has yet come forth to identify the culprit – the cause of the anomaly. Conventional physics has no explanation!

New Theories

This has turned out to be one of the most intriguing unsolved mysteries to confounded scientists for the past two decades. Theorist Michael Nieto acknowledges that the Pioneer anomaly "reveals that there might be something wrong with our understanding of gravity, the most pervasive force in the universe."[14] The Pioneer spacecraft has been the first real test of Newtonian gravity outside of the solar system, and things aren't working the way they're supposed to. Nieto states the case succinctly. "Everything about gravity is mysterious."[15]

As time lapses, new theories are being proposed to explain this peculiar phenomenon of the Pioneer spacecraft. Some of these theories must of necessity deal with the nature of gravity.

One explanation proposed by Mordehai Milgrom, a physicist of the Weizmann Institute of Science in Rehovot, Israel, is that gravity would become slightly stronger over large distances of space. That is, when the gravitational acceleration drops below a 10-billionth of a meter per second squared, then the force of gravity grows stronger. The formula for the force of acceleration then changes to:

$$F = \frac{ma^2}{a_0}$$

Where:

a_0 is a new constant relating the speed of light to the lifetime of the universe.

Let's assume the lifetime of the universe is only 11 billion years old. That would be 3.4713×10^{17} seconds. And "c" is the speed of light. Then

$$a_0 = \frac{c}{3.4713 \times 10^{17}}$$

$a_0 = 8.6362 \times 10^{-10}$ meters per second squared. This is quite close to the Pioneer anomaly of 8.74×10^{-10} meters per second squared.

This theory arouses interest because the adjusted gravitational equation of Milgrom is related to the lifetime of the universe. And we now know that gravity also is somehow related to the expansion of space.

This theory is known as Modified Newtonian Dynamics, or MOND. Milgrom may have the mechanics, (i.e. the mathematical equation) worked out to give a close approximation, but there is no justification, no theoretical explanation, as to why gravity should work that way. The search for a solution still goes on. Perhaps we can help.

Tracking the Pioneers

How do we really know where the Pioneer spacecraft really are? Their transmitters have long since stopped transmitting any signals. However, we can still bounce microwave signals off each of them and time the returning signal. We can do some simple trigonometry, throw in a little geometry, shake it up with some algebra, and presto – out comes the answer. We wish!

There's just one little tiny problem with the bouncing microwaves. When they return, there is an unexpected blue shift in the Doppler frequency. What does that mean? Remember, Hubble was able to determine the direction a galaxy was moving with respect to the Earth by analyzing the color shift in the color spectrum. A red shift meant the galaxy was moving away from us; while a blue shift meant the galaxy was moving toward us.

So what does the blue shift mean for the Pioneer spacecraft? - One of possibly two things. According to Nieto and Turyshev, "The drift can be interpreted as being due to a constant acceleration of $a_p = 8.74 \times 10^{-8}$ cm/s^2 directed *towards* the Sun."[15] This is what we see and this is what we get. But, why is it blue shifted? The blue shift should mean the Pioneer is moving toward us. It's not. That is why this was an unexpected, and very peculiar phenomenon. Can we just attribute it to the Pioneer's unexpected acceleration toward us?

The second interpretation is taken by Antonio F. Ranada of the Universidad Complutense, Madrid, Spain; who sees the blue shift meaning as "an acceleration of clocks $a_t = a_p / c$."[16] I take that to mean an acceleration of time. Ranada's sixteen page paper shows the blue shift is the result of some kind of time dilation that gives the appearance of acceleration. Ranada says, "Indeed, it is an unexpected and new effect of the dark energy."[16] Also, Ranada acknowledges that "the pioneer anomaly is an effect of the background gravitational potential that pervades all the universe and is increasing because of the expansion."[16] Simply put, Ranada explains the blue shift in terms of acceleration of time, which is the result of an expanding universe. I like that!

But wait! There is still another significant interpretation by J.L. Rosales of Universidad de Cantabria, Canto Blanco; Madrid Spain; who sees the blue shift as an expansion of spacetime.[17] A point of interest with Rosales solution is the scale of view. He sees it as a manifestation of the cosmological expansion *in the solar system.* This is a very cautious approach by Rosales, as opposed to Ranada's sweeping statement about the background gravitational potential that pervades the entire universe. Rosales relates the Pioneer effect to Foucault's pendulum. Showing the Pioneer effect based on solar system coordinates, and therefore not being the true frame of reference with respect to the expansion of the universe is very insightful. He may very well be correct, but the essential key to this whole Pioneer problem is the expanding universe.

Another important part of Rosales' solution is in the mathematical equations he presents to show why the Pioneer effect is not seen in the planets and other bodies that orbit the Sun. This may be a very significant find in the search for a true understanding of the Pioneer anomaly. Anderson and Nieto are resurrecting the telemetry data on Pioneer 11 as it traveled between Jupiter and Saturn at an angle perpendicular to the Sun.[18] Analyzing of data still goes on.

Expanding Space

There is very little we understand about expanding space and the expanding universe. What we do now know for sure is that space is expanding. Just how does an expanding universe affect gravity? In the last few years, since the discoveries made by Perlmutter and his team of astronomers, and Anderson's team to investigate the Pioneer anomaly, progress to understand the expanding universe is just beginning. Comments made by both Ranada and Rosales in their papers, clearly indicate an expanding universe.

Let's see some of their quotes again:

J.L. Rosales & J.L. Sanchez-Gomez:

- the Pioneer effect is nothing else but the detection of cosmological expansion in the solar system.
- a new effect seems to have unexpectedly entered the phenomenology of physics.

Antonio F. Ranada:

- The anomaly is thus in this model an effect of the expansion of the universe.
- Indeed, it is an unexpected and new effect of the dark energy.
- it is an effect of the increasing background gravitational potential that pervades the universe and produces an acceleration of the time...

Permit me to present a very simplistic view of the Pioneer anomaly, and what it means. The blue shift we're seeing from the microwaves bounced off the Pioneer spacecraft is a result of space expansion. The microwave signal is traveling through expanding space. As space is expanded, time is contracted. I guess that's why they call it spacetime. Whether you view it as an acceleration of the clock, time dilation, or expansion of space, one thing is for certain – the universe just got bigger.

We have no firm answer for the Pioneer anomaly yet. Let's take the unexplained acceleration of the Pioneers that has been observed by these scientists, 8.74×10^{-10} meters per second squared, and hold onto it. Perhaps we'll find an explanation later. We'll put the Pioneer anomaly on the shelf for now. We simply do not have enough information at the present time to solve this problem. In the next chapter, we are about to get into another unexplained mystery of science. The next mystery is the galaxy rotation problem. The Pioneer anomaly and the galaxy rotation problem appear to be two unrelated mysteries. We will be looking for a common thread that connects them.

Chapter 11

The Galaxy Rotation Problem

Galaxies spin like a top.

Planet's Orbital Speed

In the chapter on Kepler's Laws we discussed Kepler's Law of Periods. We saw that there was a relationship between the distance a planet is from the Sun and the time it takes a planet to orbit the Sun. This is also a relationship between a planet's distance from the Sun and the planet's velocity of orbit. A planet's distance from the Sun is inversely proportional to its velocity of orbit.

The table below shows the planets and their distance from the Sun in meters, The orbital velocity of the planets is in kilometers per second. Compare the second and third columns of the table to see the relationship between a planet's distance from the Sun and its velocity of orbit. This table shows that the farther a planet is from the Sun, the slower it will travel in its journey around the Sun.

Table 11 – Orbital Velocity of the Planets

Planet	Distance in meters	Orbital Velocity	Distance * Speed sq	Acceleration in millimeters
Mercury	5.7909E+10	47.88	1.328E+20	39.5858
Venus	1.0821E+11	35.02	1.327E+20	11.3373
Earth	1.4960E+11	29.79	1.327E+20	5.9318
Mars	2.2794E+11	24.13	1.327E+20	2.5551
Jupiter	7.7841E+11	13.07	1.330E+20	0.2191
Saturn	1.4267E+12	9.67	1.335E+20	0.0652
Uranus	2.8710E+12	6.84	1.341E+20	0.0161
Neptune	4.4983E+12	5.48	1.350E+20	0.0066

The Fundamental Force

A planet's orbital velocity can be calculated by the following equation:

Equation 16 – Orbital Velocity

$$v = \sqrt{\frac{Gm}{d}}$$

Where:

    G   is the Gravitational Constant.

    m   is the mass of the object holding the other in orbit

    d   is the distance between the two objects

    v   is the velocity of the object in orbit around the other

The radius of a planet's orbit and its orbital velocity squared is the same for each planet, as can be seen in the fourth column titled "Distance * Speed sq". The value obtained also happens to be the value of the gravitational constant multiplied by the mass of the Sun (Gm), as shown in the equation below.

Equation 17 – Gm

$$dv^2 = Gm$$

Where:

    G   is the Gravitational Constant.

    m   is the mass of the object holding the other in orbit

    d   is the distance between the two objects

    v   is the velocity of the object in orbit around the other

The acceleration of each planet towards the Sun is shown in the table's last column. The acceleration is given in millimeters per second squared. The acceleration of each planet can also be readily calculated from the planet's distance from the Sun and its velocity using the following equation.

Equation 18 – Gravitational Acceleration

$$a = \frac{v^2}{d}$$

Where:

    a   is the acceleration of the object in orbit
    v   is the velocity of the object in orbit
    d   is the distance between the two objects

Figure 11 gives a graphical picture to help visualize the relationship between the orbital speed at which each of the planets travel and its distance from the Sun in Astronomical Units.

Figure 11 – Orbital Velocities

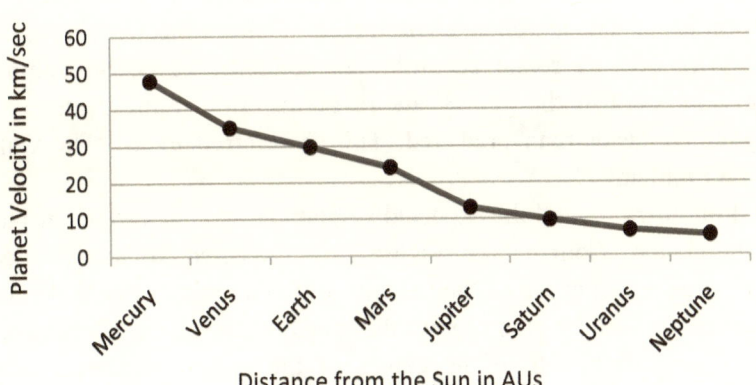

This graphic shows again that the farther away a planet is from the Sun the slower it will travel in its orbit. We know that the planet Mercury takes only 88 days to cycle the Sun; and Mercury is relatively very close to the Sun. Uranus and Neptune are much more distant and take 84 and 164 years respectively to make their orbits. The speed at which Mercury travels in orbit around the Sun is 47.87 kilometers per second. Uranus and Neptune crawl along at a relatively slow pace of 6.79 kilometers per second and 5.43 kilometers per second respectively.

This may seem intuitively obvious, but this information becomes an issue of great importance in the problem described below.

The Galaxy Rotation Problem

With all this understanding of Kepler's Laws and planetary orbits, combined with improvements in telemetry such as the Hubble space telescope, astronomers were well prepared to expand our understanding of the galaxies. New research began with our own galaxy, the Milky Way. Astronomers were able to determine the direction and speed of a star's movement in the Milky Way. By making numerous observations and putting all the data together they were able to determine the rotational speed and direction of rotation of the Milky Way itself. They were even able to determine the position of our solar system at the edge of one of the spiral arms.

This is great and exciting news, but there's a fly in the ointment. The observations of the astronomers don't show the expected results. The Milky Way doesn't rotate according to Kepler's Laws. Something is wrong. The stars in the galaxy are revolving around the center of the galaxy much faster than predicted. This is referred to as "The Galaxy Rotation Problem".

Astronomers plotted the expected velocity of stars at various distances from the center of our galaxy. Figure 12 – Expected Rotational Speeds, shows the expected curve of the velocities of stars within the Milky Way. The vertical axis of the graph shows the speed of the stars in kilometers per second. The distance from the center of the galaxy is shown on the horizontal axis in kilo parsecs. A parsec is defined as 3.262 light years. The horizontal axis spans 101,000 light years (3.262 * 31 kilo parsecs). The diameter of the Milky Way is estimated to be 80,000-100,000 light years across. That would make the radius about 45,000 light years, or about 14 kilo parsecs. That is thought to be the extent of the visible part of the Milky Way.

The Milky Way is a spiral galaxy with a bar at the core. The bar (and core) of the Milky Way is estimated to be about 8 kilo parsecs (26,000 light years) in diameter. Stars at the edge of the core were expected to have the greatest velocity. Figure 12 shows that at the edge of the core, 4 kilo parsecs from center, the star speeds are at their peak. At that distance they are traveling at speeds of approximately 220 kilometers per second. This graph shows that stars beyond the galactic core (4 kilo parsecs) should gradually decrease in speed as they get further away from the galaxies' center. This would be as predicted from Kepler's Laws.

Figure 12 – Expected Rotational Speeds

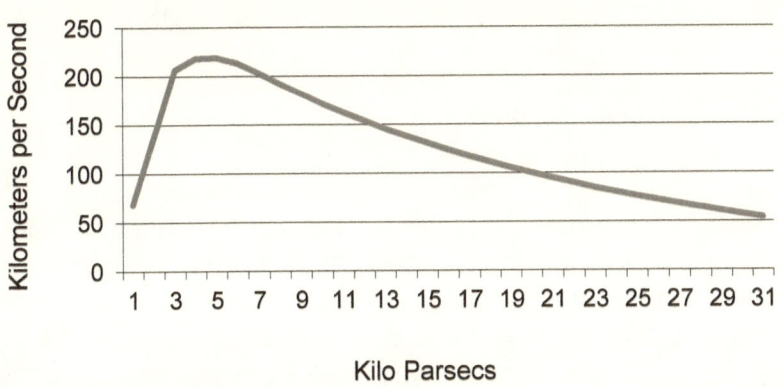

Kilo Parsecs

Figure 13 – Actual Rotational Speeds, shows the actual observed speeds of revolution of the stars around the galactic center. This is significantly different than expected. After the peak speed is reached the curve levels out. It only decreases with distance ever so slightly. This indicates that the galaxy is spinning like a top – almost like a solid object.

Figure 13 – Actual Rotational Speeds

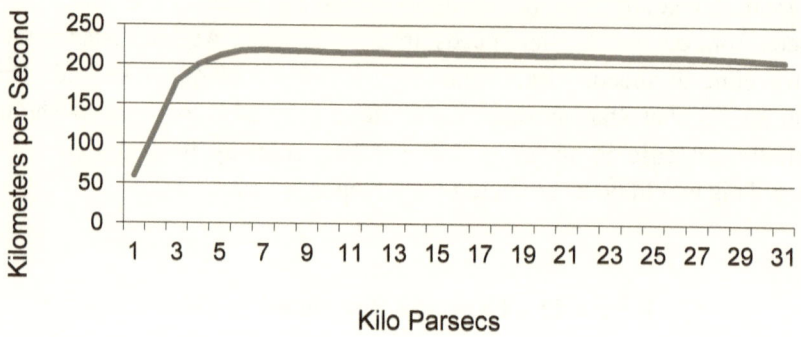

The Galaxy Rotation Dilemma

Spiral galaxies do not rotate as expected according to the laws described by Kepler. For most people this may not be a cause for concern and will be dismissed as a minor anomaly to be resolved later. Scientist's on the other hand are at work trying to solve this dilemma.

One explanation suggests that there is a substantial amount of matter circling the outer fringes (halo) of the galaxies. Some believe that this extra mass keeps the stars in the galaxy rotating at a higher than expected speed. The problem with this theory is that the matter cannot be seen. Thus it is called "Dark Matter." Astronomers are now searching for the evidence of dark matter. Because it cannot be seen, new techniques must be developed to measure its effects. A few astronomers believe they have witnessed these effects in the interactions between galaxies.

One speculative explanation as to why this matter cannot be seen proposes that the matter in the halo of the galaxies did not form into large enough clumps, which then weren't large enough to ignite into stars. With this matter not giving off light, it would be difficult to see. Therefore there could possibly be a large amount of matter in small chunks of various sizes. The sizes could range from dust specks to larger bodies like Jupiter. As long as they are not large enough to ignite, they could be considered dark matter. Since this kind of matter is not burning like a star nor giving off light of its own, some refer to it as "cold dark matter."

We could compare this with the planetoids within the Kuiper belt of our solar system. The amount of matter in the belt relative to the solar system is insignificant. Some suggest that dark matter comprises more of the matter in the galaxy than the regular kind of visible matter.

Another explanation for why we can't see dark matter is that dark matter is not made out of the fundamental particles of normal matter, i.e. electrons and protons. The speculation is that dark matter is made from something else; some other exotic particles. This may be why some call it "exotic dark matter."

In any case, there is much speculation and confusion about the existence and nature of dark matter. Some suggest that 75 to 90 percent of the universe is composed of dark matter. Some say dark matter has anti-gravitational properties and is the cause of the expansion of the universe. Still others say that dark matter does not interact with regular matter or itself. There has been quite a lot of articles written and speculation on something that may not even exist.

Many physicists and astronomers are not buying into the dark matter controversy. Some are looking into alternative explanations of the galactic rotation problem. One interesting alternative is given by Mordehai Milgrom of the Weizmann Institute of Science in Rehovot, Israel. Milgrom suggests that gravity actually gets stronger over large distances. This theory is referred to as Modified Newtonian Dynamics, or MOND. The MOND theory was previously mentioned in the chapter on The Pioneer Anomaly.

Ockham's Razor

A principle referred to as Ockham's Razor is often used by scientists when multiple hypotheses, theories, or speculations are raised to explain some unusual physical phenomena. Such is the case with dark matter. The principle is attributed to William of Ockham (1285-1349), an English theologian and philosopher. The principle translated from Latin states, "entities should not be multiplied unnecessarily."[19]

This principle advises us that new entities such as MOND or dark matter should not be prescribed as solutions if the phenomenon can be explained otherwise. At the time when the galaxy rotation problem was first discovered, and the discrepancy with Kepler's law revealed; there was no explanation other than that of dark matter. Since the entity of dark matter is in a state of confusion, it's time to revisit the galaxy rotation problem.

In our analysis of the solar system and Kepler's Laws, we made the assumption that the mass of the planets was insignificant compared to the mass of the Sun. This was a good assumption, and our numbers for the Law of Periods were quite accurate. But suppose the Sun were much smaller in mass than it actually is. In that case the masses of the planets would affect the calculations, and the resulting numbers would be less accurate.

So it was with the original assumptions made by scientists concerning the structure of the galaxies. It was originally believed that most of the mass in a galaxy was at its core. This is where the big bulge is, where the density of matter is supposedly the greatest, and where a massive black hole is expected to reside. So it was logical to assume that most of mass was in the core and calculations were done accordingly as was done for the solar system. The estimated mass of the core was used in the calculations while the rest of the mass was disregarded as insignificant. The resulting calculations show the expected rotational speeds of the stars in Figure 12.

The graph for the galaxy in Figure 12 is very similar to the one for the solar system in Figure 11, in that velocity decreases with distance. This would be expected because all of the assumptions were similar. However, the actual observed data as shown in graphic form in Figure 13, indicates that those assumptions were not true. The velocities of the stars reach a peak and stay there. This means that from the peak on, the stars all travel in their orbits at about the same speed. The observed data is telling us what the structure of the galaxy really is. We just have to understand what's happening. Do we need to re-adjust our assumptions to fit the observed data? Or do we need to create other entities to explain this phenomenon?

Attempted Dilemma Resolution

In my own efforts to resolve this dilemma, I built a spread sheet model of the Milky Way based on the information currently available. I used a line in the spread sheet for each kilo-parsec of space from the center of the galaxy out to the edge. I positioned our solar system 8 kilo-parsecs from center, just like the astronomers at Hayden Planetarium. I modeled the shape as a typical spiral galaxy making the core wider than the disk. I spread the mass evenly throughout the disk from the edge of the core to the edge of the galaxy.

Newton and Kepler's equations were used to calculate the vital statistics. I calculated the acceleration of the stars at each kilo-parsec. I also calculated the speed of their orbits. Necessary adjustments were made to the amount of mass in the core and disk in order to get the data for our solar system up to the correct magnitude.

To my own amazement, the rotational speed curve of my model matched with the actual rotational speed curve obtained by exhaustive measurement. The values for the solar system's velocity and time of rotation around the galaxy are right on target. Our solar system is traveling in orbit about the center of the Milky Way at a speed of about 217 kilometers per second. At a distance of 8 kilo-parsecs from center, it would take over 226 million years to complete just one cycle. It takes 1.74×10^{41} kilograms of mass in the central area of the Milky Way in order for our solar system to travel at that speed. No matter what shape the core of the Milky Way really is, it takes 1.74×10^{41} kilograms of mass in the area of a sphere that would enclose the central area of the galaxy, from its center out 8 kilo-parsecs in every direction.

Quite delighted with myself I just sat back to relax - job well done. After a few brief moments of self exultation I thought I'd better review the accuracy of my spread sheet. The modeled rotational velocities of the stars match up well with the measured curve. The rotational speed of our solar system matched the stated values of the astronomers. Even the total mass of the galaxy was within the range specified by the scientists. Everything was right on target but there was still something wrong.

The scientists investigating the nature of the universe and our galaxy are pretty smart people. They've already done these calculations, obtaining the same results. So then what conclusions can be drawn from this little spread sheet experiment? It would seem Ockham's Razor applies! There seems to be no need for new entities like dark matter, or any modification to the laws of gravity. Everything appears to work quite well according to the laws of Newton and Kepler. It looks like our galaxy can be modeled according to the existing laws of physics. Everything is fine if the mass is spread out evenly throughout the galaxy and not condensed at the center, as was previously thought. But, there's still something very wrong.

The Spiral Arm Problem

So if everything matches, what's the problem? The problem is in the spiral arms – they exist. Some current estimates say that forty percent of the galaxies in the universe are spirals. There are spiral arms in the Milky Way. So what's the problem? That can't happen given the data we have just confirmed. If the stars in a galaxy are revolving around the center at about the same speed, spiral arms would exist only for a very short time at the beginning of a galaxy's life. After just a few cycles the arms would blend in and all galaxies would be lenticular.

Here's why. The stars in the region 4 kilo-parsecs from the center of a galaxy would be traveling at about the same speed as those which are 8 kilo-parsecs from center. The stars 8 kilo-parsecs from center are twice the distance from center as those that are 4 kilo-parsecs off center. This means that their circumference is twice as large. This also means that stars twice as far from center take twice as long to make a complete orbit. In a galaxy with arms, after several cycles there would be just a homogenous blend of stars with no discernable spiral arms at all. Any spiral arms in a galaxy would be short lived. But, spiral arms do exist in great abundance throughout the galaxies in the universe.

Some scientists speculate the existence of density waves to explain the spiral arms. The origin of density waves is unknown. How they can propagate through space and create spiral arms is also unknown. Let's apply Ockham's Razor to this for now and see if we can better explain the existence of spiral arms.

Some scientists suggest that when galaxies are forming with very little angular momentum they will form into elliptical galaxies. No spin – no spiral. When a forming galaxy has a high amount of angular momentum (lots of spin), they will form into a spiral galaxy. I like that – it makes sense. The ends of the ellipse become the spiral arms; the middle of the ellipse becomes the bar of the spiral. This is an excellent concept but the problem is that the spiral arms will all blend in after a few cycles.

The Milky Way is believed to be 13.6 billion years old and it still has its spiral arms. Andromeda is believed to be about the same age and it still has its spiral arms. Why? I took a copy of my handy dandy spread sheet and made another galactic model that would keep its spiral arms over a long term period. All I needed to do was increase the angular momentum of the stars that were further from center. So, as I doubled the distance, I doubled the velocity. By doing this a galaxy could keep its spiral arms for billions of years. That's really neat, but comes with a high price tag.

In order to get those stars up to speed, more mass had to be added to the galaxy. The mass increased in the area for each kilo-parsec further away from center. This meant that the mass on the outer edge of the galactic disk was greater than on the inner part of the disk. But it just doesn't look that way. It must be dark matter, the kind you can't see. And there must be ten times more of it than of the real visible matter.

Now I understand the problem. This is why scientists are looking for dark matter. If we don't like the dark matter theory we have another option. We still might reconsider Mordehai Milgrom's MOND theory. That's the Modified Newtonian Dynamics theory that suggests that gravity gets stronger over large distances. I didn't buy that at first but it's beginning to look like a good idea right now. It looks like these are the only two choices available, and both are being actively pursued by scientists.

Speed is the Issue

Before presenting an alternate solution, a problem needs to be pointed out and addressed. The problem was just introduced when attempting to preserve the galactic arms by increasing the velocity of the stars as their distance from the center of the galaxy increases. The velocities of the stars have been measured and are all very close to being the same. This has been depicted in the graphic showing a flat curve in Figure 13 – Actual Rotational Speeds. The astronomers are all in agreement on this although some slight variations may exist. Let me address this problem with the following analogy.

Let's assume we and several other scientists are aboard a large jet airliner. We have all the latest scientific equipment with us and the task at hand is to measure the motion of a ball rolling down the fuselage of this airplane in flight. The scientists are divided up into teams with each team using their own testing tools, and taking their own measurements as they see fit. After the ball rolling is complete each team tabulates their scores. All teams present their results to the committee in charge of score validation to evaluate their scores. What was the velocity of the ball rolled down the aisle, and what was its acceleration, etc?

After the judges review the results, they come to the conclusion that all teams came up with the same results, and all were correct. Any small differences of velocity or acceleration could be attributed to equipment sensitivity or adjustments. Essentially all teams reported the same results. But we know that all motion is relative. What was the velocity of the ball to another scientist who was also observing, but not on the airplane? That would be an extremely difficult thing to measure. Perhaps that's why those measurements have not been made. But, those results would be very different from all the teams on the jet airliner.

So it is with the velocity of the stars revolving around the center of the galaxy. We are within the galaxy, taking measurements of stars also within the galaxy. The complete motion of the galaxy had not actually been measured in these experiments. So, there's still a piece of the puzzle missing!

Chapter 12

Galactic Rotation

There's more to galactic rotation than meets the eye.

Galactic Rotation and Expanding Space

The solution to the galaxy rotation problem can be explained by the expansion of space. We understand now that the universe is expanding, and as it does, the space within it also expands. We have no way of quantifying this expansion except through the measurable effects it has on objects in space. On a smaller scale, such as our solar system, these effects are not easily observed. However, the effect of expanding space on a galactic scale is much more pronounced. Envision space expanding in all directions outward from the center of our galaxy. The effect this expansion has on the objects in the galaxy has never been understood until now.

When we focus on any particular object in space, it appears that the expansion of space is emanating from the object. This is diametrically opposite of gravity. With gravity, objects of mass are contracting space. With space expansion, space is expanding out from objects of mass. This is why we should envision the space within our galaxy as expanding in all directions from the center of the galaxy.

It may appear odd that objects can be contracting space while at the same time, space is expanding out of those objects. How can that be? A detailed explanation of how space expansion works will be forthcoming along with supporting evidence. For now, once again, let's briefly consider the units of measurement of the Gravitational Constant.

$$G = \frac{\text{meters}^3}{\text{kilogram} * \text{second}^2}$$

We previously suggested that we can interpret the Gravitational Constant to be a measurement of a specific volume of space being contracted toward, and consumed by one kilogram of mass. We viewed the Gravitational Constant as a rate of contraction and consumption of space based on a specific amount of matter. On the other hand, we can also view it as the rate of space expansion. The rate of space expansion is also based on kilograms of mass. If that is true, then it's easy to visualize space expanding out from matter, just as space is contracted toward matter.

Consider again the units of measurement of the expression GM:

$$GM = \frac{meters^3}{second^2}$$

This is a measurement of the rate of change of volume of space. We will see that the expression GM is used for both space contraction and space expansion. For now let's assume this to be true; more explanation and evidence will be presented in following chapters.

Acceleration via Expansion

There are two effects that cause acceleration of bodies in space. The first is gravity. Gravity is the contraction of space which has the powerful effect of accelerating bodies in space toward the source of contraction; this is easily observed and measured. The second is the expansion of space. But this acceleration is not *outward*, aligned with the expansion; rather it is *inward*, toward the source of expansion. Just like gravity. In both cases, the acceleration is toward some massive object or objects.

In the case of gravity it is simple to understand why an object will accelerate toward the source of contracting space. Space is actually contracting toward every object; therefore every object attracts every other object causing an acceleration toward it. So stars anywhere in a galaxy will accelerate toward the source of contraction – the mass at the center of the galaxy.

In the case of expanding space, the space between a star and the center of mass in the galaxy is being extended beyond the star in a direct line connecting the center of the star and the center of the galaxy. This has an effect on the star that "feels like" the space between the star and the center of the galaxy has been lessened. It's an apparent shortening of space. This is what increases the angular momentum of the rotating stars and results in an increase of velocity.

Let's use the typical explanation given to describe the law of angular momentum. An ice skater plants the point of the blade of her skate into the ice to act as the pivot. Then she energetically swings her other leg around with her arms extended to start a spin. As she brings her leg and arms in close to her body she begins to spin much faster. Her spin rate increased in order to preserve angular momentum.

So it is with the expansion of space. As space expands around a galaxy, it's *as though* the stars got closer to the center even though their distance from the center actually remains the same. The spin rate increases in order to preserve angular momentum. That is what appears to happen, but a galaxy is slightly different than an ice skater. In the case of a galaxy the expansion of space does not actually change the distance of the stars from the center of the galaxy, but causes the angular momentum of the stars to be increased *as though* the space were shortened. With this additional angular momentum the acceleration of the stars increase, and so does their velocity.

What we have are two different phenomena working in diametrically opposite directions but having a similar affect on objects. Gravity is the contraction of space caused by mass. Therefore gravity has the effect on all objects of attracting all other objects, from all directions. Gravity is an inward attraction of mass toward mass. The expansion of space works in the opposite direction. As space expands outward from massive objects it passes through other objects in the vicinity. The reaction of these objects to the expansion is acceleration toward the source of expansion. It's as if the space between the object and the source of expansion had been shortened. That results in the same effect that gravity has – acceleration toward the source of contraction or in this case, the source of the expansion.

This means that there are two different phenomena that cause acceleration. The acceleration caused by gravity is simply the effect of contracting space. The acceleration caused by expanding space gives added angular momentum to the stars in the galaxies, which dramatically increases their velocities. And this is the phenomenon that makes gravity appear to grow stronger over large distances. It's no wonder that Mordehai Milgrom came up with the theory of Modified Newtonian Dynamics. Milgrom certainly had the right idea even though he did not have a theoretical basis for his theory. Milgrom now has a foundation for his theory with more explanation yet to come.

Why Galaxies Rotate

Galaxies rotate for two reasons. The first is the gravitational effects of contracting space. This is the one we all know about. We can measure the speed of stars and calculate the time of rotation. But now we know there's more to it than just that. The second reason is the acceleration resulting from expanding space. This will cause the galaxy to rotate as a whole. It is a rotation that goes undetected and can only be seen and measured from outside the galaxy.

We can liken this rotation to a sink filled with water. Pull out the stopper and watch what happens. The water begins to drain; at the same time it begins to spin. Look directly down the drain and you'll see a black hole. Flush a toilet and the same thing happens. The spin of the falling water in the northern hemisphere of the Earth is in a counter clockwise direction. In the southern hemisphere the rotation is clockwise. A very slight imbalance across the water caused by the rotation of the Earth is enough to start the spin.

Galaxies rotate for the same reason. All matter in the galaxy is falling toward the center of attraction. A very slight imbalance across the galaxy will start the spin. Space is continually expanding around the galaxy. The effect of the expansion is to increase the angular momentum of all the matter within the galaxy. This increases the velocity of all the objects in orbit around the galaxy's center. Thus the spin picks up speed. All the matter in the galaxy spins around its center. However, the falling never brings it any closer to the center of the galaxy once equilibrium is reached. All the matter falls just like the planets do in orbit around the Sun. All the stars in a galaxy will revolve around its center, never getting any closer when all is balanced. Thus a galaxy's rotation is powered by the expansion of space. This also explains why galaxies rotate in nearly circular orbits.

Scientists believe that there is a massive black hole in the center of our galaxy, and many others. The black hole may act as the pivot point of rotation. Black holes have been referred to as the anchor of the galaxies.

One other point needs to be mentioned. There is no force applied to initiate the rotation of the objects within galaxies. Just as in the example of the ice skater, once her arms and leg were drawn in, her spin rate increased. Angular momentum is increased with no force applied. The effect of both gravity and expansion is the same – acceleration. The acceleration is a consequence of falling. Everything is falling! If there is any force at all it would be the force associated with expanding space – dark energy.

Once again, galaxies rotate for two reasons, expanding space and gravity. The rotation caused by gravity is according to the laws of Kepler and Newton. The rotation caused by expanding space is a new phenomenon of physics that is just now being understood. Both are working in concert to give the galaxies spin and velocity. The combined effect makes the spiral galaxies rotate as if they were solid. If we were to look down from space high above the Milky Way and speed up time, we would see it rotating as a whole. The spiral arms would still be intact after 13.6 billion years. What a magnificent sight that would be!

Rotation Velocity

Let's quantify the effect that expanding space has on matter to get a better understanding of the rotating galaxy phenomenon. It appears that spiral galaxies rotate almost like a solid object. In order for this to happen the velocity of stars must increase in direct proportion to their distance from the center of the galactic mass. That is exactly what is happening in galaxies. The velocity of the stars double as the distance from the galactic center doubles.

Table 12 – Galaxy Rotation, is a partial listing of a spread sheet used to analyze the rotation of the Milky Way. "Kpc" is the number of Kilo-parsecs from the center of the galaxy. This listing only goes up to 15 Kpcs. The limit of the visible galaxy would end at 14 or 15 kilo-parsecs. One kilo-parsec (1 Kpc) is a light year times 3,262; which is equal to $3.0857e^{19}$ meters. That's a long distance by any stretch of the imagination. The

The Fundamental Force

radius at each kilo-parsec is shown in meters in the second column - entitled "Radius".

Table 12 – Galaxy Rotation

Kpc	2*Kpc	Radius	Velocity	Velocity X	Kpc mass	Mass
1	2	3.086E+19	179,627	359,255	1.492E+40	1.492E+40
2	4	6.171E+19	206,689	826,754	2.458E+40	3.950E+40
3	6	9.257E+19	217,243	1,303,456	2.596E+40	6.546E+40
4	8	1.234E+20	217,152	1,737,215	2.175E+40	8.720E+40
5	10	1.543E+20	217,097	2,170,974	2.175E+40	1.090E+41
6	12	1.851E+20	217,061	2,604,732	2.175E+40	1.307E+41
7	14	2.160E+20	217,035	3,038,491	2.175E+40	1.524E+41
8	16	2.469E+20	217,016	3,472,249	2.175E+40	1.742E+41
9	18	2.777E+20	217,000	3,906,008	2.175E+40	1.959E+41
10	20	3.086E+20	216,988	4,339,767	2.175E+40	2.177E+41
11	22	3.394E+20	216,978	4,773,525	2.175E+40	2.394E+41
12	24	3.703E+20	216,970	5,207,284	2.175E+40	2.612E+41
13	26	4.011E+20	216,963	5,641,042	2.175E+40	2.829E+41
14	28	4.320E+20	216,957	6,074,801	2.175E+40	3.047E+41
15	30	4.629E+20	216,952	6,508,559	2.175E+40	3.264E+41

The velocity of the stars at the Kpc distance from the center of the galaxy is given in meters per second and is shown in the "Velocity" column. These velocities are calculated by the equation below which was derived from Kepler and Newton's laws. These velocities match up very well with astronomical observations. These are the velocities being observed from within the galaxy.

Equation 19 – Gravitational Velocity

$$v = \sqrt{\frac{Gm}{d}}$$

Where:

- v   is the velocity of objects
- G   is the Gravitational Constant
- m   is the mass from which the attraction emanates
- d   is the distance between the object and the center of attraction

The following graph shows the velocities from the "Velocity" column in Table 12.

Figure 14 – Rotational Velocity

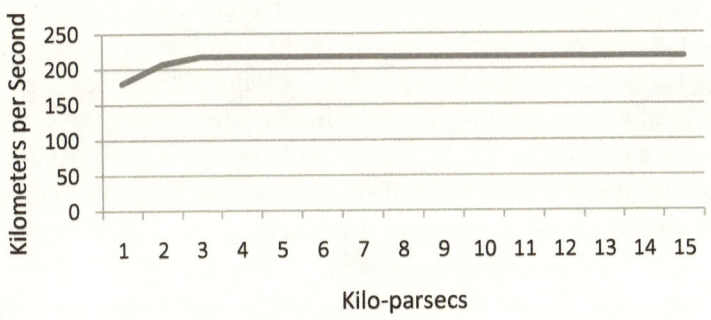

The velocities in the column entitled, 'Velocity X', are those calculated with expanding space. The "$2*K_{pc}$" column is the number in the "K_{pc}" column multiplied by 2. Notice that ($2 * K_{pc}$) times the velocity in the "Velocity" column is equal to the velocity in the "Velocity X" column. Here we can see a very clear relationship between the normal velocity from gravitational acceleration and the velocity from expanding space. This relationship is shown in the equation below:

Equation 20 – Velocity via Space Expansion

$$V_x = 2K_{pc} * V$$

Where:

V_x is the velocity of objects with space expansion included

V is the velocity of the object from just gravity

K_{pc} is the number of kilo-parsecs from the source of expansion

As can be seen from Table 12, the velocities from expanding space (Velocity X), get dramatically higher with distance. Row 15 shows that Velocity X is 30 times greater than the normal Velocity (30 * 216,952 = 6,508,560). This calculation doesn't exactly match the table because the numbers in the spread sheet from which the table was derived contain decimals places whereas our calculations use no decimals.

The "K_{pc} mass" column shows the amount of mass contained in just that area within that kilo-parsec. The area for each K_{pc} is defined by the sphere with the K_{pc} radius minus the area of the next lower order K_{pc}. Each of the K_{pc} masses are added to the "Mass" column. The "Mass" column shows the total mass at each K_{pc} distance from center of the galaxy. Notice that the mass contained in the "K_{pc} mass" column is the same from row 4 on. This shows that the mass in the disk of the galaxy is evenly spread out within the galactic arms. In the past many scientists assumed incorrectly that most of the galactic mass was centered in the core.

A visual check can be deceptive. We only see the stars in pictures. Most of the mass may still be in the form of hydrogen dust between the stars. There is so much hydrogen dust that it blocks out our view of the galactic core. Looking at the night sky, we do not see the galactic core.

Matter is spread out evenly throughout the galactic disk. When this is taken into account Kepler's laws still work as advertised. The expansion of space and the increase of angular momentum it causes were previously unaccounted for. This additional momentum causes the stars to revolve around the center of the galaxy much faster than expected. Although this has given the appearance that the galaxies are not rotating according to Kepler's Laws, those laws continue to work as expected.

Comparing Velocities

Figure 15 – Comparing Velocities, illustrates the difference in the velocities attributed to gravity and space expansion. The span of space in this comparison is from the center of the Milky Way out to a distance of 31 kilo-parsecs, or about 100,000 light years. The vertical axis shows the velocity of stars in millions of meters per second. The horizontal axis across the bottom of the graph represents the distance in Kilo-parsecs from the center of the galaxy.

Figure 15 – Comparing Velocities

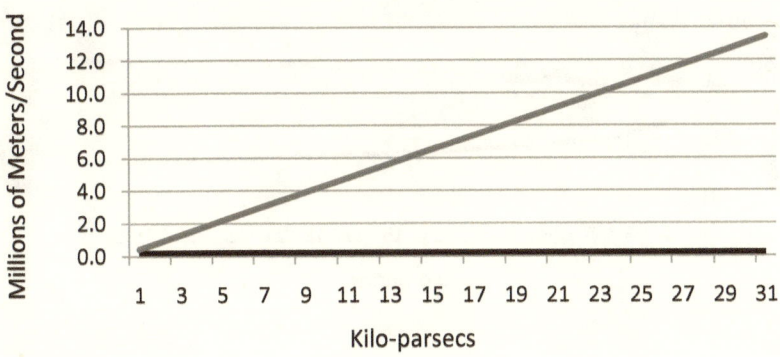

The black line across the bottom of the graph represents the velocity of stars attributed to gravitational acceleration. This velocity is nearly constant at about 217,000 meters per second. This is what we see in the "Velocity" column of Table 12. This velocity can be attributed to the effects of gravity.

The inclining gray line represents the velocity of stars increasing the further they are from the center of the galaxy. This is the real velocity of the stars which can be attributed to the effects of both gravity and space expansion. Once again, this velocity is not being measured by scientists, because we are all inside the rotating galaxy.

This graph shows a flat line for the velocity due to the effects of gravity, and a line of inclination due to the effects of expanding space. The reason the lines are as they are is because we have modeled the galaxy with its mass evenly distributed throughout. If there were no significant mass other than within the first kilo-parsec, then the graph would appear as shown in Figure 16.

Figure 16 – Comparing Velocities

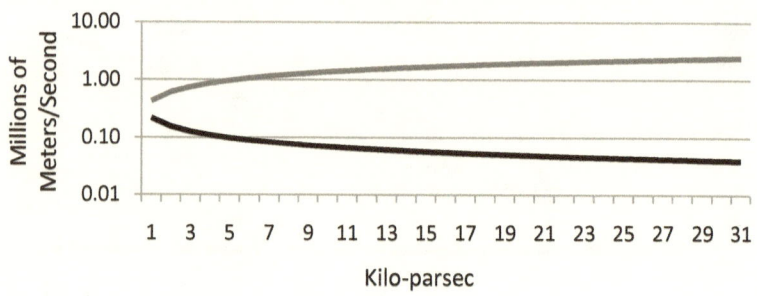

Figure 16 is a logarithmic graph. The velocities appear to be a mirror image of each other. The velocity due to gravity declines as we get further from the center of the galaxy. The velocity due to space expansion increases as we get further from the galactic core. The interesting thing is that the rates of inclination and declination appear to be the same, only in opposite directions. This would lead one to believe that gravity and space expansion are intimately related – opposite sides of the same coin.

Grasp this picture! Gravity is the contraction of space. Space expansion refers to the expansion of space. The expansion of space is always centered on objects of mass, and the contraction of space is always toward objects of mass. Space expansion occurs everywhere, but appears to be centered on objects of mass and expanding out from the mass. There are two different, but related, phenomena working on all objects of mass in the universe. Mass contracts the space around it, and space expands out from those objects.

Galaxy Rotation

Now let's examine Table 13 – Galaxy Rotation. The column for Rotations gives the number of times the stars in the specified K_{pc} have rotated around the center of the galaxy, in the whole lifetime of the galaxy. This column was calculated by the normal velocity that scientist have observed. In the 8th K_{pc} we can see why our solar system is believed to have made only 60 rotations. Our solar system resides close to 8 K_{pc} from galactic center. The slot in the 8th row of the Rotations column shows that our solar system has made about 60 rotations in its lifetime. This table also contains a column titled "Gravity X" which will be explained in detail in the following section.

Table 13 – Galaxy Rotation

Kpc	a	ax	Gravity X	Rotations	RX
1	1.046E-09	4.183E-09	2.670E-10	397.635	795.270
2	6.922E-10	1.108E-08	1.068E-09	228.770	915.079
3	5.098E-10	1.835E-08	2.403E-09	160.301	961.806
4	3.820E-10	2.445E-08	4.271E-09	120.176	961.404
5	3.055E-10	3.055E-08	6.674E-09	96.116	961.163
6	2.545E-10	3.665E-08	9.611E-09	80.084	961.002
7	2.181E-10	4.274E-08	1.308E-08	68.635	960.887
8	1.908E-10	4.884E-08	1.709E-08	60.050	960.801
9	1.696E-10	5.494E-08	2.162E-08	53.374	960.734
10	1.526E-10	6.104E-08	2.670E-08	48.034	960.680
11	1.387E-10	6.713E-08	3.230E-08	43.665	960.636
12	1.271E-10	7.323E-08	3.844E-08	40.025	960.600
13	1.173E-10	7.933E-08	4.512E-08	36.945	960.569
14	1.090E-10	8.543E-08	5.233E-08	34.305	960.542
15	1.017E-10	9.152E-08	6.007E-08	32.017	960.519

The "RX" column provides valuable insight into the way the Milky Way, and all galaxies, rotate. This column is based on the velocities of the stars calculated with space expansion included. The "RX" column shows the number of rotations that stars in the specified K_{pc} have made in the lifetime of the galaxy. This column is very interesting because it shows the Milky Way is a spiral galaxy. The extremely small drop in rotations with distance is exactly what would be expected for spiral galaxies. The number of rotations in the 3rd row of the "RX" column is 961.806. This row in the table depicts the area 3 K_{pc} from center. This would be about at the edge of the galactic core. The number of rotations in the 15th row is 960.519. The 15th row depicts the edge of the visible galactic disk, at the tip of a spiral arm.

The Fundamental Force

Figure 17 – The Milky Way

Image Credit: NASA/JPL-Caltech

http://www.nasa.gov/mission_pages/spitzer/multimedia/20080603a.html

Let's do some simple arithmetic to illustrate the point. Subtract 960.519 from 961.806 and get 1.287 rotations. The Milky Way is believed to be 13.6 billion years old. So in the lifetime of the Milky Way, the outer edge of the disk has lagged behind the core by only 1.287 turns. This is fairly close to what we're seeing in the pictures of galaxies. Looking at an artist's depiction of the Milky Way, (Figure 17), we can only see about 1 turn.

The numbers in the "RX" column can be easily adjusted as will be seen shortly. Figure 18, shows a graph of the "RX" column. This graph shows the rotations as a straight line from the 3rd kilo-parsec on to the edge of the galaxy. Since the stars in these sectors throughout the galaxy are all making about the same number of rotations, the galaxy appears to be rotating almost as a solid object. Stars further away from center are in a slightly slower rotation which causes the spiral.

Figure 18 – Number of Rotations

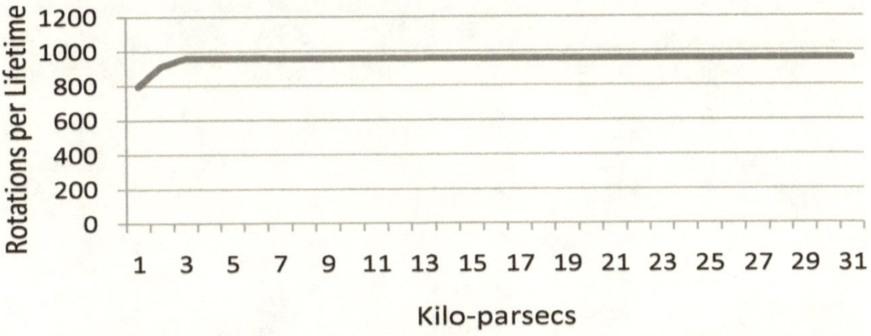

The Figure 19 is of spiral galaxy NGC 4622. What is of interest here is that it appears to be spinning backwards.[20] This is very puzzling to astronomers. Why should this galaxy, and perhaps a few others just like it, be rotating backwards? Actually this galaxy is spinning in a clockwise direction with the spiral arms leading and pointing the way instead of trailing behind. Although this seems very peculiar, it may actually be a common occurrence. We can achieve that effect by making one simple little adjustment to our model.

Figure 19 – Spiral Galaxy NGC 4622

Image Credit: NASA and The Hubble Heritage Team[21]

Acknowledgment: Dr. Ron Buta (U. Alabama), Dr. Gene Byrd (U. Alabama) and Tarsh Freeman (Bevill State Community College)

Table 14 – Galaxy Rotation in Reverse, shows selected columns from Table 12 and Table 13. This was the model of the Milky Way. There is only one little change that needs to be made to one of the slots in this table to change the characteristics of its rotation. The amount of mass in the first row of the "nK$_{pc}$ mass" column has been reduced from 1.492e40 to 1.450e40, a reduction in mass of less than 3 percent. The other rows of this column remain the same.

Table 14 – Galaxy Rotation in Reverse

Kpc	nKpc mass	Mass	RX
1	1.450E+40	1.450E+40	784.063
2	2.458E+40	3.908E+40	910.231
3	2.596E+40	6.504E+40	958.734
4	2.175E+40	8.679E+40	959.100
5	2.175E+40	1.085E+41	959.320
6	2.175E+40	1.303E+41	959.466
7	2.175E+40	1.520E+41	959.570
8	2.175E+40	1.738E+41	959.649
9	2.175E+40	1.955E+41	959.710
10	2.175E+40	2.173E+41	959.759
11	2.175E+40	2.390E+41	959.798
12	2.175E+40	2.608E+41	959.832
13	2.175E+40	2.825E+41	959.860
14	2.175E+40	3.043E+41	959.884
15	2.175E+40	3.260E+41	959.905

This one modification has a ripple effect through the other columns in the table. Each row of the "Mass" column is slightly reduced. But, the "RX" column now shows an interesting effect. Remember the "RX" column shows the number of revolutions that stars in the specified K$_{pc}$ range have made. Each successive row now slightly increases instead of decreases. This of course affects the appearance of the galaxy. Instead of having spiral arms that lag behind, the spiral arms are leading the way. This would give the appearance that the galaxy is rotating backwards.

This appearance of reverse rotation would indicate that the NGC 4622 spiral galaxy has a smaller than usual amount of mass in its central core. When viewing Figure 19 – Spiral Galaxy NGC 4622, visualize the galaxy rotating in a clockwise direction.

This little exercise also illustrates that the characteristics that form and shape a galaxy can be finely tuned by making slight adjustments to the amount of mass in and around the galactic core.

All of this is very convincing evidence that gives credence to the supposition that space expansion is the explanation for galactic spin. But hold off judgment for just a little while longer. There's much more evidence to uncover.

Dark Matter and MOND

We have previously seen the relationship between normal velocity and the velocity with space expansion included. The normal velocity was calculated with the equation below, which comes from Kepler's laws.

Equation 21 – Velocity

$$v = \sqrt{\frac{Gm}{d}}$$

Where:
- v   is the velocity of objects
- G   is the Gravitational Constant
- m   is the mass from which the attraction emanates
- d   is the distance between the object and the center of attraction

The velocity of objects with space expansion included is calculated the same way but is multiplied by 2 times the number of K_{pcs} from the center of the expansion. That would be the same location as the center of attraction.

Equation 22 – Velocity via Space Expansion

$$v_x = 2K_{pc} \sqrt{\frac{Gm}{d}}$$

Where:
- v_x is the velocity of objects with space expansion included

K_{pc} is the number of kilo-parsecs from the source of expansion

G   is the Gravitational Constant

m   is the mass from which the expansion/contraction emanates

d   is the distance between the object and center of expansion

Let's make a slight readjustment to this equation by including the $2K_{pc}$ under the square root sign. This is done only as an aid to explanation. The equation then looks like this:

Equation 23 – Velocity via Space Expansion

$$v_x = \sqrt{\frac{4K_{pc}^2 * Gm}{d}}$$

This equation explains the increase of gravitational attraction that the proponents of the MOND theory have been looking for. The column entitled 'Gravity X' in Table 13 shows the strength of the gravitational constant at the K_{pc} distances from galactic center. This equation can also be used to explain the dark matter that scientists have been searching for.

Scientists have been bewildered by the high velocities of stars on the outer perimeter of the galaxy. Some have even suggested that they are traveling so fast that they should have been flung off into outer space beyond the galaxy. Two theories have been invoked to explain this peculiar phenomenon. The two prominent theories are Dark Matter and the Modification of Newtonian Dynamics – MOND. The proponents of the Dark Matter theory say there must be missing mass, mass that cannot be seen, which holds the stars in orbit. The proponents of MOND suggest that the gravitational force grows stronger over large distances.

The equation for velocity above shows both cases to be kind of true. First let's consider the missing mass. The missing mass is found! At 1 kilo-parsec distance from the center of the galaxy the expression $4 K_{pc}^2 * GM$ becomes 4GM. At only 1 kilo-parsec from center, galaxy mass has an apparent fourfold increase. At 2 kilo-parsecs the expression $4 K_{pc}^2 * GM$ becomes 16GM. At 2 kilo-parsec from center, galaxy mass appears to increase by a factor of 16. At 3 kilo-parsecs mass appears to increase by

36; and so on. So of course it would appear that galaxies have much more mass than they really do.

Looking at this from the MOND prospective, the strength of the gravitational force is 4 times stronger at 1 kilo-parsec from center galaxy. At 2 kilo-parsecs from center, the gravitation force appears to becomes 16 times stronger than at 1 kilo-parsec. The strength of gravity appears to increase with distance. This is exactly what Milgrom, the founder of MOND, was expecting; an increase in gravitational strength over distance. Previously there was no justification for this increase; but now there is a comprehensive explanation.

Table 13 – Galaxy Rotation, contains a column entitled "Gravity X". That column shows the value of the gravitational constant at each kilo-parsec from center. Actually that would make the gravitational constant not a constant. So this column needs clarification. The value of the gravitational constant does *not* actually change as the MOND theory would suggest. This is just a convenient way to do calculations for objects from the center of the galaxy. Multiplying the gravitational constant by 4 K_{pc}^2 at the K_{pc} distance, all calculations for accelerations and velocities could be done using the normal Newtonian equations.

More Spin on the Galaxies

The following modifications to our spread sheet are being made to illustrate the effects that small adjustments, to the mass in the galactic core, have on galactic rotation. Table 15 is a slightly modified version of Table 12. In this case the "K_{pc} mass 1" column is modified so that rows 2 and 3 would contain the same amount of mass as the other rows lower down in the column. The amount of mass removed from columns 2 and 3 was added to row 1. This shifting of mass kept the observed velocity of our solar system (in row 8) over 217,000 meters per second. The resulting number of rotations is shown in the "Rotations 1" column.

Table 15 – Galaxy Rotation

Kpc	Velocity	Kpc Mass 1	Rotations 1	Kpc Mass 2	Rotations 2
1	217,836	2.1939E+40	964.433	2.1752E+40	960.314
2	217,371	2.1752E+40	962.376	2.1752E+40	960.314
3	217,216	2.1752E+40	961.689	2.1752E+40	960.314
4	217,139	2.1752E+40	961.346	2.1752E+40	960.314
5	217,092	2.1752E+40	961.139	2.1752E+40	960.314
6	217,061	2.1752E+40	961.002	2.1752E+40	960.314
7	217,039	2.1752E+40	960.904	2.1752E+40	960.314
8	217,022	2.1752E+40	960.830	2.1752E+40	960.314
9	217,009	2.1752E+40	960.773	2.1752E+40	960.314
10	216,999	2.1752E+40	960.727	2.1752E+40	960.314
11	216,990	2.1752E+40	960.690	2.1752E+40	960.314
12	216,983	2.1752E+40	960.658	2.1752E+40	960.314
13	216,977	2.1752E+40	960.632	2.1752E+40	960.314
14	216,972	2.1752E+40	960.609	2.1752E+40	960.314
15	216,968	2.1752E+40	960.589	2.1752E+40	960.314

Comparing the "Rotations 1" column in this table with the "RX" column of Table 12 illustrates the significance of the location of mass in the galactic core. A small relocation of mass in the core of a galaxy affects the number of rotations in the spiral arms. Calculating the difference in rotations by subtracting the number of rotations in row 15 of the "Rotations 1" column (960.589) from the number of rotations in row 3 (961.689) gives 1.1, which is just slightly more than one full turn. This same calculation from Table 13 gave a turn difference of 1.28 turns.

The column "K_{pc} mass 2" is the same as column "K_{pc} mass 1" except for the first row. In column "K_{pc} mass 2" all the rows have the same amount of mass. The resulting numbers for rotations in the column "Rotations 2" are all the same. So when the mass is evenly spread out throughout a galaxy, it will spin like a solid object.

This tells us that all those elliptical galaxies scattered throughout the universe are still elliptical because their mass is spread evenly throughout. If the galaxy had developed with just a little more mass in the core, it would have become a spiral galaxy. This table also tells us that elliptical galaxies are rotating just as rapidly as spiral galaxies, and they are rotating as a solid object.

The argument that elliptical galaxies do not have enough angular momentum to develop into spirals does not stand up to reason. If they lacked angular momentum they would have collapsed in on themselves. This of course would have resulted in incredibly huge explosions, putting supernova to shame. This also would have left super massive black holes where once there was an elliptical galaxy.

Chapter 13

Space Expansion

Space expands like a rubber band.

Expanding Space

How does expanding space work? Many scientists have illustrated the concept of an expanding universe by using a rubber band. Put several dots on a rubber band to represent galaxies. Then stretch the rubber band and watch them expand away from each other at the same rate. This analogy also works quite well with a balloon. Put dots on a balloon to represent galaxies. As the balloon is blown up the dots move away from each other. Likewise, supposedly, the galaxies are carried along with the expanding space. If we take this assumption to be true, then we would have to say that objects tend to follow the path of expanding space.

Some scientists suggest that objects of mass are repelled from each other via expanding space. This is not what's happening. In the rubber band and balloon analogies described above the dots represent galaxies. Yes, they all receded from each other, but the dots also got bigger. If objects tended to follow the path of expansion, then the galaxies would continually grow larger and eventually fall apart. There is no such evidence of anything like that ever happening. No galaxy has ever just fallen apart. Likewise, our solar system would forever expand away from the Sun. This is not happening, therefore let's reject this analogy and the hypothesis that objects are being carried along by the expansion of space – they are not. Let's also remember that the Pioneer spacecraft are not moving along with expanding space. If they were, there would have been no blue shift in the color spectrum, and no Pioneer anomaly. The unaccounted for acceleration of the Pioneer spacecraft is toward the Sun, not away from the Sun. Space expands like a rubber band, but does not carry objects with it.

The Fundamental Force

How then does expanding space work? Expanding space gives additional acceleration to objects in space. That acceleration is toward the source of expansion. There's more explanation on acceleration via expanding space in the following sections.

Acceleration via Expanding Space

In the previous chapter we looked at the velocities of stars within galaxies and the rotational velocities of those galaxies. In this chapter we'll examine the acceleration caused by the expansion of space.

Table 16 – Accelerations in the Galaxy, is a continuation of tables from the previous chapter. It differs by showing accelerations of stars within a galaxy rather than velocities.

Table 16 – Accelerations in the Galaxy

Kpc	2*Kpc	Radius	a	ax	Kpc mass	Mass
1	2	3.086E+19	1.046E-09	4.183E-09	1.492E+40	1.492E+40
2	4	6.171E+19	6.922E-10	1.108E-08	2.458E+40	3.950E+40
3	6	9.257E+19	5.098E-10	1.835E-08	2.596E+40	6.546E+40
4	8	1.234E+20	3.820E-10	2.445E-08	2.175E+40	8.720E+40
5	10	1.543E+20	3.055E-10	3.055E-08	2.175E+40	1.090E+41
6	12	1.851E+20	2.545E-10	3.665E-08	2.175E+40	1.307E+41
7	14	2.160E+20	2.181E-10	4.274E-08	2.175E+40	1.524E+41
8	16	2.469E+20	1.908E-10	4.884E-08	2.175E+40	1.742E+41
9	18	2.777E+20	1.696E-10	5.494E-08	2.175E+40	1.959E+41
10	20	3.086E+20	1.526E-10	6.104E-08	2.175E+40	2.177E+41
11	22	3.394E+20	1.387E-10	6.713E-08	2.175E+40	2.394E+41
12	24	3.703E+20	1.271E-10	7.323E-08	2.175E+40	2.612E+41
13	26	4.011E+20	1.173E-10	7.933E-08	2.175E+40	2.829E+41
14	28	4.320E+20	1.090E-10	8.543E-08	2.175E+40	3.047E+41
15	30	4.629E+20	1.017E-10	9.152E-08	2.175E+40	3.264E+41

Column "a" is the acceleration rate of stars at each K_{pc}. This is calculated using Equation 6 – Acceleration. This is the normal acceleration on mass due to the gravitational effect. Column "ax" shows the acceleration rate on stars due to the combined effects of gravity and space expansion.

A brief comparison of the accelerations reveals an interesting phenomenon. The acceleration due to gravity weakens over distance, decreasing according to the inverse square law. The results in column "ax", the combined acceleration of both effects, increases over distance. The relationship between the two accelerations is shown in the equation below. This is very similar to what we had seen with the Gravitational Constant in the previous chapter.

Equation 24 – Acceleration via Expanding Space

$$a_x = (2K_{pc})^2 * a$$

Where:
- a_x is the acceleration rate of objects with space expansion
- a is the acceleration rate of objects from just gravity
- K_{pc} is the number of kilo-parsecs from the source of expansion

The equation below is the one we've been using all along to calculate the acceleration of objects due to gravity.

$$a = \frac{GM}{d^2}$$

We can replace the acceleration (a) in Equation 24 – Acceleration via Expanding Space, with its equivalent expression to get a new equation. This new equation, shown below, can be used to calculate the combined acceleration of objects due to the effects of gravity and expanding space at the specified K_{pc} distance.

Equation 25 – Acceleration via Expanding Space

$$a_x = \frac{(2K_{pc})^2 \, GM}{d^2}$$

Where:
- a_x is the acceleration of objects with space expansion included
- K_{pc} is the number of kilo-parsecs from the source of expansion
- G is the Gravitational Constant

M   is the mass from which the expansion/contraction emanates
d   is the distance between the object and center of expansion

Expansion Acceleration Rate

Let's modify the equation again so we have the distance in terms of the number of kilo-parsecs (K_{pc}) and the size of a kilo-parsec (kpc). It's basically the same equation expressed in a different way.

$$a_x = \frac{4\,(K_{pc})^2 * GM}{(K_{pc})^2 * kpc^2}$$

The reason for making this change is so that we can see that the expression $(K_{pc})^2$ can be factored out of the equation; leaving us with the equation in its final form.

Equation 26 – Acceleration via Expanding Space

$$a_x = \frac{4GM}{kpc^2}$$

Where:
    a_x is the acceleration of objects with space expansion included
    G   is the Gravitational Constant
    M   is the mass from which the expansion emanates
    kpc is the length of 1 kilo-parsec in meters – 3.0857×10^{19} meters

Factoring $(K_{pc})^2$ out of the equation shows that the expression for acceleration is a linear function. Everything is a constant except for mass. The mass is not constant because it increases with each K_{pc} from the galactic center. This equation would only be valid for galaxies whose mass is evenly distributed and for distances greater than 1 kilo-parsec from center.

Figure 20 – Combined Accelerations shows the expansion acceleration in graphic form. In this case the expansion acceleration came from the "ax" column of Table 16 – Accelerations in the Galaxy. Acceleration increases with each additional kilo-parsec because the mass of the galaxy also increases with each kilo-parsec.

Figure 20 – Combined Accelerations

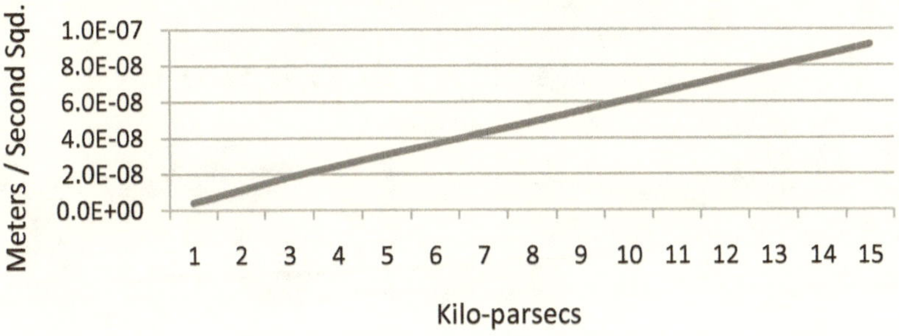

Figure 20 shows the acceleration across 15 kilo-parsecs of the Milky Way, (which is the visible part of the galaxy), with matter spread evenly throughout. This illustrates the fact that the acceleration of stars increase with distance from the center of the galaxy.

Comparing Accelerations Above 1 Kilo Parsec

At 1 kilo-parsec from the galactic center the acceleration resulting from gravity is only ¼ of the expansion acceleration. Beyond that, the acceleration from gravity becomes much less significant. Compare the accelerations from Table 16. At 15 kilo-parsecs from center of the Milky Way the acceleration caused by space expansion is 900 times greater than the gravitational acceleration. Beyond 1 kilo-parsec, gravitational acceleration becomes insignificant, thus the acceleration of stars resulting from space expansion dominates the expanse of the galaxy which causes the rapid rotation.

The graph for expansion acceleration shown in Figure 20, shows the combined acceleration of gravity and space expansion. Figure 21 shows both accelerations on a logarithmic scale. Once again, as with velocities, we have a mirror image. As gravitational acceleration declines with

The Fundamental Force

distance, acceleration due to expanding space inclines with distance. They appear to be opposite sides of the same coin. This is assuming that the mass is spread evenly throughout the galactic disk. Also notice that these accelerations will be crossing paths below 1 K_{pc}.

Figure 21 – Accelerations Compared

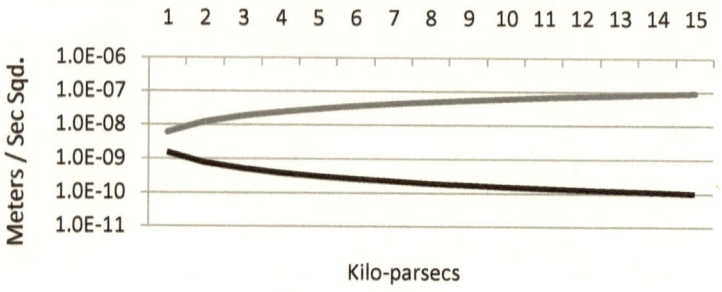

Kilo-parsecs

Expansion From One Central Source

If all the matter in the galaxy existed within only the first kilo-parsec then the graph would look considerably different. Table 17 is a modified version of Table 16. Table 17 has been modified so that there is no increase of mass beyond the first K_{pc}. The "Kpc mass" column shows no increase of mass beyond 1 K_{pc}. The "Mass" column now shows the mass to be constant. The reason for making this hypothetical adjustment is to show the effect that expanding space has on objects from just one central body of mass. In this case the central body of mass has all the matter at the core of the Milky Way; within a radius of 1 K_{pc}.

Table 17 – Accelerations from the Galactic Core

Kpc	2*Kpc	Radius	a	ax	Kpc mass	Mass
1	2	3.086E+19	1.046E-09	4.183E-09	1.492E+40	1.492E+40
2	4	6.171E+19	2.614E-10	4.183E-09	0.000E+00	1.492E+40
3	6	9.257E+19	1.162E-10	4.183E-09	0.000E+00	1.492E+40
4	8	1.234E+20	6.535E-11	4.183E-09	0.000E+00	1.492E+40
5	10	1.543E+20	4.183E-11	4.183E-09	0.000E+00	1.492E+40
6	12	1.851E+20	2.905E-11	4.183E-09	0.000E+00	1.492E+40
7	14	2.160E+20	2.134E-11	4.183E-09	0.000E+00	1.492E+40
8	16	2.469E+20	1.634E-11	4.183E-09	0.000E+00	1.492E+40
9	18	2.777E+20	1.291E-11	4.183E-09	0.000E+00	1.492E+40
10	20	3.086E+20	1.046E-11	4.183E-09	0.000E+00	1.492E+40
11	22	3.394E+20	8.642E-12	4.183E-09	0.000E+00	1.492E+40
12	24	3.703E+20	7.262E-12	4.183E-09	0.000E+00	1.492E+40
13	26	4.011E+20	6.187E-12	4.183E-09	0.000E+00	1.492E+40
14	28	4.320E+20	5.335E-12	4.183E-09	0.000E+00	1.492E+40
15	30	4.629E+20	4.647E-12	4.183E-09	0.000E+00	1.492E+40

The total mass of 1.492×10^{40} kilograms for this hypothetical galaxy comes from the first row of the "Mass" column. All of the mass is contained within the first K_{pc}. The "ax" column shows the acceleration rate to be perfectly flat. Objects would accelerate at the same rate across the expanse of the galaxy because space is expanding at a steady rate. The mass from which space appears to be expanding from, would be centered in the galactic core.

Figure 22 illustrates the steady acceleration resulting from expanding space.

Figure 22 – Combined Accelerations

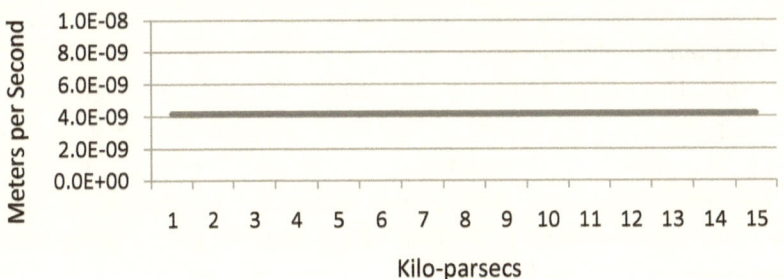

This graph is showing the "ax" column from Table 17, which is actually the combined acceleration from contracting and expanding space. The "a" column shows the gravitational acceleration to be very small and is declining with distance. The gravitational acceleration becomes insignificant with distance.

The Rubber Band Rule

The combined accelerations can be likened to a stretching rubber band. Take a rubber band and cut it so it has ends. Lay it flat next to a ruler without any tension on it. Mark the rubber band at the 1, 2, and 3 inch lines of the ruler. Hold one end of the rubber band down at the edge of the ruler, as shown in the following diagram:

Figure 23 – Rubber Band Rule 1

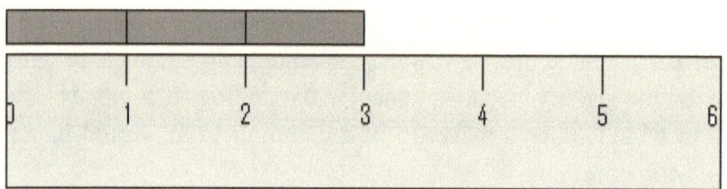

Stretch the other end of the rubber band so that the mark corresponding to the 3 inch line will now be lined up with the 6 inch line of the ruler. The mark for the 1 inch line is now at the 2 inch line of the ruler. The mark for the 2 inch line will now be lined up with the 4 inch line of the ruler. The rubber band stretches evenly, but the further away from the fixed point, the anchor, the greater the expansion appears to be. The mark on the rubber band that started at the one inch line only traveled one inch. The mark that started at the 3 inch line traveled three inches. Space expands like a rubber band. In the case of galaxies, the anchor is the center of mass in the galactic core.

Figure 24 – Rubber Band Rule 2

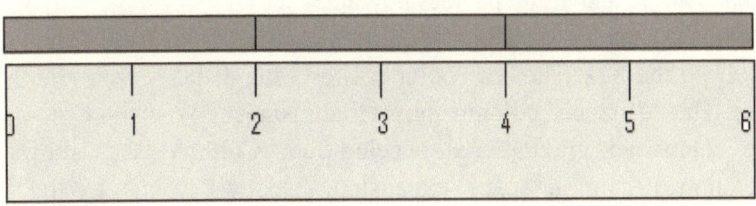

Just one more experiment to illustrate the nature of expanding space. Take a cut rubber band and make a mark every inch from the end. Hold the rubber band out by its ends without stretching it. Focus on the mark nearest the center, and slowly stretch it out. It's very easy to visualize the center mark as being the anchor point when the ends are being stretched out. Do the same thing again, only this time focus on one of the marks off center. The rubber band still stretches the same way, evenly in all directions. Any point could be the center of expanding space. However, since space expansion and gravity are intimately connected, and space is being contracted toward the center of mass, the center of mass needs to be the anchor point in our calculations.

One other very important point needs to be made. The rubber band analogy is only an analogy that breaks down at some point. A rubber band stretched beyond its capacity will eventually break; that is not the case with space expansion. The expansion of space goes on with the expansion of the universe with no dilution. As space expands it always maintains its essential characteristics; so the characteristics of space remain the same throughout the universe as the universe expands. The same can be said of dark energy. As the universe expands the dark energy within is not diluted.

Comparing Accelerations Below 1 Kilo Parsec

Let's bring the accelerations down to a smaller scale. Instead of using a hypothetical galaxy let's use the Sun and the surrounding space to make our calculations. One kilo-parsec is 1000 parsecs. Let's compare the accelerations of expanding space and contracting space across 1000 parsecs. These 1,000 parsecs represent the distance from our Sun out into space. So, we have significantly downsized the scale from the galaxy to our solar system and 1,000 parsecs beyond.

Table 18 shows the difference in accelerations from 1000 parsecs to 1 parsec from the Sun. The "pc" column shows the distance from the Sun in parsecs. The "distance" column shows that distance in meters. Column "a" is the calculated gravitational acceleration. Column "ax" shows the acceleration just from space expansion. This allows us to track the differences in accelerations as we get closer to the Sun. Note that both accelerations increase as we get closer to the Sun, but the gravitational acceleration increases much faster than the acceleration due to space expansion.

Table 18 – Acceleration from Sun in Parsecs

pc	distance	a	ax
1000	3.0857E+19	1.3942E-19	4.1827E-19
900	2.7771E+19	1.7213E-19	4.4925E-19
800	2.4685E+19	2.1785E-19	4.8798E-19
700	2.1600E+19	2.8454E-19	5.3777E-19
600	1.8514E+19	3.8728E-19	6.0416E-19
500	1.5428E+19	5.5769E-19	6.9711E-19
400	1.2343E+19	8.7139E-19	8.3654E-19
300	9.2570E+18	1.5491E-18	1.0689E-18
200	6.1714E+18	3.4856E-18	1.5336E-18
100	3.0857E+18	1.3942E-17	2.9279E-18
10	3.0857E+17	1.3942E-15	2.8024E-17
1	3.0857E+16	1.3942E-13	2.7898E-16

It is much easier to understand the balance of accelerations by looking at a graphic picture that shows their relationship. Figure 25 does just that. The previous graphs in this chapter show the distance from the center of the galaxy increasing as we traverse from left to right. In the following graphs the distance decreases from left to right.

Figure 25 shows the curve for each of the accelerations. The curve for gravitational acceleration is black while the curve for space expansion acceleration is gray. At 1,000 parsecs from the Sun the gravitational acceleration is extremely weak, only 1.39×10^{-19} m/s^2. At this distance the acceleration from space expansion is also extremely weak, but 3 times stronger at $4.18\text{x}10^{-19}$ m/s^2. The accelerations get closer with each 100 parsecs, and converge at about the 400 parsec marker. At that point the accelerations are about equal. As the number of parsecs decreases the gravitational acceleration begins to rapidly increase. The space expansion acceleration also increases but much less dramatically. At 1 parsec from the Sun the gravitational acceleration is at $1.39\text{x}10^{-13}$, while the acceleration from space expansion is only a meager $2.78\text{x}10^{-16}$; about 500 times weaker.

Figure 25 – Parsec Accelerations Compared

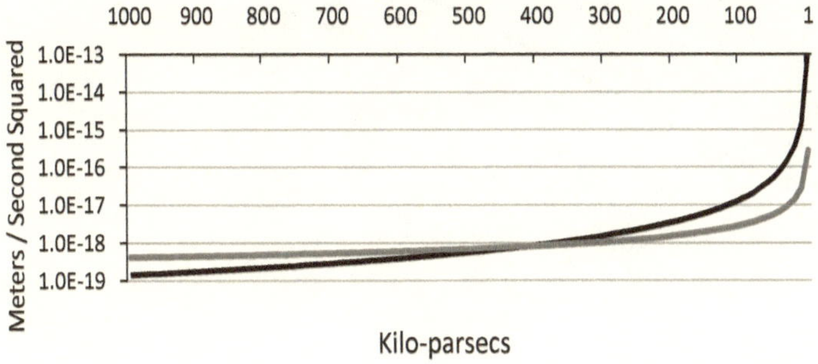

Comparing Accelerations in Parsecs

Figure 26 is essentially the same chart but in a different format. This logarithmic graph shows both rates of accelerations from 1000 parsecs to just 1 billionth of a parsec. This is accomplished by successively reducing the distance by a factor of 10. The starting point horizontally is at marker 1 on the graph, and is at a distance of 1,000 parsecs from the Sun. One tenth of that distance is 100 parsecs, at marker number 2. It's between these two markers where the lines representing the two accelerations cross. The third marker is 10 parsecs from the Sun.

By continually reducing the size by one tenth, we will be at one billionth of a parsec (3.0857e^7 meters) at the 13th marker at the right side of the chart. The 13th marker is over 30 million meters from the Sun. This chart shows both accelerations on the incline as they approach the Sun.

The Fundamental Force

However, the gravitational acceleration is increasing much faster than that from space expansion. The gap widens as we get closer to the Sun.

Figure 26 – Accelerations Compared

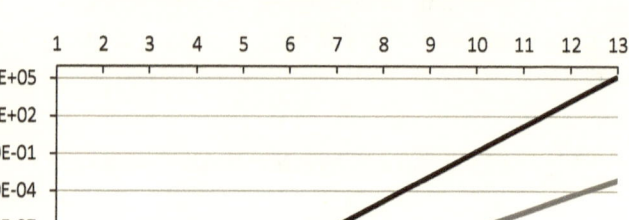

From 1000 Parsecs to 1 billionth of a Parsec

Comparing Accelerations in Astronomical Units

Let's reduce the scale again to view these accelerations in more detail in Astronomical Units. This time we'll use Figure 27 to chart from 100 Astronomical Units from the Sun to 1. One Astronomical Unit from the Sun is the radius of the Earth's orbit around the Sun. This is a rather severe reduction of scale because there are 206,264 Astronomical Units in 1 parsec. We will be charting only the last 100 of those Astronomical Units. These are the most interesting because they are nearest the Sun.

Figure 27 – Accelerations Compared

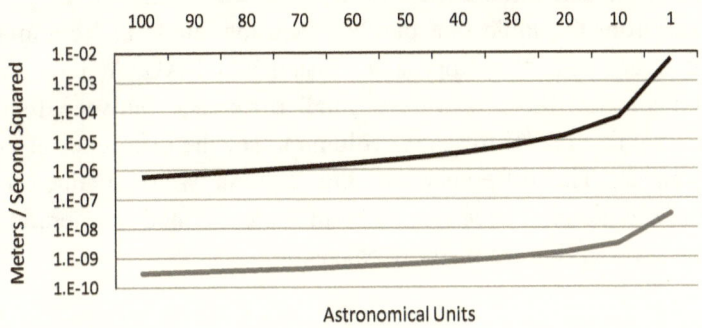

The black line on the chart represents the normal gravitational acceleration. The gray line below represents acceleration from space expansion. Both increase as the distance gets closer to the Sun, however these lines are not parallel. The rate of acceleration due to space expansion lags behind. This is a logarithmic chart so the difference between the horizontal axis lines is a factor of 10. At 100 AUs from the Sun the gravitational acceleration is almost 2,000 times greater than space expansion acceleration. The gravitational acceleration at 1 Astronomical Unit is almost 200,000 times greater than the acceleration from space expansion. The closer we get to the Sun, the greater the difference in accelerations. Which means, gravity rules inner space.

Calculating Expansion Acceleration

The last diagram presented in Figure 27, which compared the acceleration rates from 100 Astronomical Units to 1, was derived from the data from the modeling spread sheet, shown in Table 19. The column "AUs" shows the number of Astronomical Units from the Sun. The "distance" column shows the distance from the Sun in meters. The "acceleration" column shows the gravitational acceleration as calculated from the acceleration equation. The "ax" column shows the combined accelerations from gravity and space expansion. The "difference" column

The Fundamental Force

shows the difference between the "acceleration" column and the "ax" column. This is the acceleration attributed to just space expansion.

We can see from the table that the "acceleration" and "ax" columns appear to be converging as they approach a distance of 1 AU. At 1 AU we see no difference. But, there is a very small difference between these accelerations. Examine the "difference" column to see the difference at the various AU ranges. The differences in this column were obtained by subtracting the gravitational acceleration from the combined acceleration, leaving just the acceleration from space expansion.

Table 19 – Expansion Acceleration

AUs	distance	acceleration	ax	difference	X accel
100	1.496E+13	5.9316E-07	5.9346E-07	3.0295E-10	3.0295E-10
90	1.346E+13	7.3229E-07	7.3263E-07	3.3661E-10	3.3661E-10
80	1.197E+13	9.2681E-07	9.2719E-07	3.7869E-10	3.7869E-10
70	1.047E+13	1.2105E-06	1.2110E-06	4.3279E-10	4.3279E-10
60	8.976E+12	1.6477E-06	1.6482E-06	5.0492E-10	5.0492E-10
50	7.480E+12	2.3726E-06	2.3732E-06	6.0590E-10	6.0590E-10
40	5.984E+12	3.7072E-06	3.7080E-06	7.5738E-10	7.5738E-10
30	4.488E+12	6.5907E-06	6.5917E-06	1.0098E-09	1.0098E-09
20	2.992E+12	1.4829E-05	1.4830E-05	1.5148E-09	1.5148E-09
10	1.496E+12	5.9316E-05	5.9319E-05	3.0295E-09	3.0295E-09
1	1.496E+11	5.9316E-03	5.9316E-03	3.0295E-08	3.0295E-08

Let's compare the accelerations. At 100 AUs from the Sun the gravitational acceleration toward the Sun is $5.9316e^{-7}$ meters per second squared, while space expansion acceleration is only $3.0295e^{-10}$ meters per second squared. At that distance gravitational acceleration is almost 2,000 times greater than space expansion acceleration.

At 1 AU from the Sun, we already know what the gravitational acceleration is. It's $5.9316e^{-3}$ meters per seconds squared, or 5.9316 millimeters per second squared. We had already calculated that acceleration in the chapters on Gravity. That's the gravitational acceleration of the Earth toward the Sun. The combined acceleration from gravity and space expansion is still only 5.9316 millimeters per second squared. The space expansion acceleration at 1 Astronomical Unit, in the "difference" column, is only $3.0295e^{-8}$ meters per second squared. This is almost 200,000 times less than the acceleration from gravity.

The acceleration from space expansion is so much smaller than the gravitational acceleration that it kind of "gets lost". This is why scientists have not seen the effects of acceleration by space expansion. In inner space this effect is simply too small to make any measurable difference in our calculations. All accelerations calculated by scientists are automatically attributed to gravity. In inner space gravity has the dominant effect. Beyond a few hundred parsecs, the effects of space expansion begin to take over. Beyond about 400 parsecs, space expansion begins to dominate.

Notice the "X accel" column of Table 19 is the same as the difference column. The "X accel" column is the space expansion acceleration calculated by the following equation. This equation is a simple way to avoid complicated math; and with it we can calculate a very good approximation of the acceleration from expanding space within the range of 100 Astronomical Units from the Sun:

Equation 27 – Acceleration via Expanding Space

$$a_x = \frac{GM}{pc_x * d}$$

Where:

- a_x is the acceleration due to expanding space.
- G is the Gravitational Constant
- M is the Mass from which the expansion emanates
- d is the distance between the object and the center of expansion
- pc_x is the length of a short parsec – 2.9291×10^{16} meters

The value of what I called a short parsec (pc_x) is 2.9291×10^{16} meters. A parsec is about 3.0867×10^{16} meters. So a short parsec is about 95 percent of a parsec, hence the name "short parsec".

Now let's use this equation to calculate the amount of acceleration on the Earth due to the effect of space expansion from the Sun. The Earth is a distance of 1 AU (1.496×10^{11} meters) from the Sun.

$$a_x = \frac{6.6742 \times 10^{-11} * 1.989 \times 10^{30}}{2.9291 \times 10^{16} * 1.496 \times 10^{11}}$$

$a_x = 3.0295 \times 10^{-8}$ meters per second squared.

Notice that the calculated acceleration matches with the acceleration in the last row of the "X accel" column of Table 19 and the "difference" column as well. The last row represents the data at 1 AU from the Sun.

Tracking the Pioneer Spacecraft

Now that we can readily calculate the acceleration for expanding space let's do another calculation. This time let's use a distance of 34.66 Astronomical units. 34.66 AUs is 5.1851×10^{12} meters from the Sun.

$$a_x = \frac{GM}{pc_x * d} = \frac{6.6742 \times 10^{-11} * 1.989 \times 10^{30}}{2.9291 \times 10^{16} * 5.1851 \times 10^{12}}$$

$a_x = 8.74 \times 10^{-10}$ meters per second squared.

Hum! That looks a little bit like a number we've seen before. Ah yes, remember the Pioneer anomaly? This number matches the unaccounted for acceleration of the Pioneer spacecraft observed by astronomers. The unaccounted for acceleration of the Pioneer spacecraft is due to the acceleration from the expansion of space.

The distance of 34.66 AUs would represent an average distance for the spacecraft. The "X accel" column of Table 19 above shows the acceleration due to space expansion across 10 AU intervals. The "acceleration" column shows the acceleration from space contraction. This table shows both accelerations slightly decreasing with distance.

Astronomers tracking the Pioneer spacecraft while they were within the solar system found no anomalous behavior. The reason this anomaly went undetected is because at short range distances, the difference in accelerations between space contraction and space expansion is large. The shorter the range of distance the greater the disparity of accelerations. At close range the acceleration from space contraction is significantly larger than the acceleration from space expansion. At a distance of one AU the acceleration from space contraction is almost 200,000 times greater than the acceleration from space expansion. So the acceleration from space expansion goes unnoticed.

At a distance of 100 AUs, the acceleration from space contraction is only about 2,000 times greater than the acceleration from space expansion. At much larger distances, the acceleration from space expansion takes over.

Table 20 below shows the accelerations from 30 to 40 AUs. The acceleration from expanding space explains the unaccounted for acceleration of the Pioneer spacecraft. This additional acceleration would decline according to these tables as the Pioneer spacecraft venture further out into space. More accurate observations of spacecraft in deep space in the future will show these accelerations to be quite accurate. This would clearly give evidence of the expansion of space.

Table 20 – Expansion Acceleration for AUs 30-40

AUs	distance	acceleration	ax	difference	X accel
40	5.984E+12	3.7072E-06	3.7080E-06	7.5738E-10	7.5738E-10
39	5.834E+12	3.8998E-06	3.9006E-06	7.7680E-10	7.7680E-10
38	5.685E+12	4.1077E-06	4.1085E-06	7.9724E-10	7.9724E-10
37	5.535E+12	4.3328E-06	4.3336E-06	8.1878E-10	8.1878E-10
36	5.386E+12	4.5768E-06	4.5777E-06	8.4153E-10	8.4153E-10
35	5.236E+12	4.8421E-06	4.8430E-06	8.6557E-10	8.6557E-10
34	5.086E+12	5.1311E-06	5.1320E-06	8.9103E-10	8.9103E-10
33	4.937E+12	5.4468E-06	5.4477E-06	9.1803E-10	9.1803E-10
32	4.787E+12	5.7926E-06	5.7935E-06	9.4672E-10	9.4672E-10
31	4.638E+12	6.1723E-06	6.1733E-06	9.7726E-10	9.7726E-10
30	4.488E+12	6.5907E-06	6.5917E-06	1.0098E-09	1.0098E-09

The understanding of space expansion has removed the shroud of mystery from two unexplained phenomena of science, the Pioneer anomaly, and the galaxy rotation problem. It also explains why the search for dark matter is no longer necessary, and gives credence to the MOND theory. Several pieces have been added to the puzzle to clarify the picture. As mentioned in the preface, there would be one discovery after another. Read on, there are more to come.

Chapter 14

Misunderstanding Tides

The tides come and go.

The Tides and Gravity

Galileo may have been the first to attempt a scientific theory of the tides. He used his explanation in an attempt to prove the Earth did in fact orbit the Sun.[22] But his tide explanation didn't wash. According to Galileo's theory there would have been only one high tide every 24 hours. Galileo wouldn't buy Kepler's explanation that the tides were caused by the moon, not the Earth. This is not an attempt in any way to denigrate a great man, but to show even men of great genius can be wrong about some things. Also, men of great genius do not necessarily agree on everything – especially very complex issues.

The cause of tides is a very complex issue with multiple explanations. Which, if any, would be the correct one? Have you ever noticed that there is no "Law of Tides", and that there are very few equations concerning tides? Some articles on tides provide explanations but show no calculations. What does all that tell us? Do we really know the cause of tides?

There are several meanings to the word "tide". For those who live by the sea shore, the meaning that may first come to mind is the rise and fall of the sea level on the coast. The costal tides vary greatly in different locations depending on the topography, so they are of little concern here. The definition of the word "tide" that we are concerned with is the tide that causes the distortion of the Earth; the tide that causes the Earth to bulge on opposite sides. What we will be investigating in this chapter is the effect gravity has on tides. How does gravity cause the tides?

Our understanding of gravity as explained by Newton states that all particles of mass attract all other particles of mass according to the inverse square of their distance. If the moon's gravity is attracting the freely maneuvering particles (water) on the surface of the Earth to itself, then all the water should bulge to one side of the Earth – nearest the moon. Therefore, simple logic would dictate only one high tide every 24 hours on the side of the Earth nearest to the moon. This seems to be a logical way to look at the problem - except it doesn't work that way. There are high tides on opposite sides of the Earth. Perhaps the cause of the tides has a much different explanation than what we currently think.

Our objective here is to better understand gravity. If we can understand tides, then we will have succeeded. One of the most interesting phenomenon of tides is the fact that we get a high tide on opposite sides of the Earth at the same time. This is an extremely difficult mystery to untangle. Any explanation of the cause of tides must explain this phenomenon.

In the chapter on Gravity, we looked at gravity in a different light. We saw gravity as an effect not a force. We will do likewise in this chapter. Now the question is, how can we possibly explain tides if gravity is not a force?

The Shape of Objects in Space

Before getting into the details of tides it would be good to understand why all large objects in space are spherical. Have we ever seen pictures from space showing a cubical planet? How about a planet shaped like a pyramid or cone? Are there any rectangular planets and cylindrical stars? No – none! There may be a lot of little space junk: comets, asteroids, small moons that come in a variety of shapes, but not in a variety of sizes. Anything with a funny shape is relatively small - too small to be classified as a planet. Any object in space large enough to be a planet is spherical.

Examine Figure 28. This simple diagram shows why large objects take on a spherical shape. The arrows show the direction of space contraction. The combined contraction of space due to all the particles of mass draws all the particles together into the smallest shape possible. All particles are drawn toward the center of the combined mass.

Figure 28 – Shaped by Space Contraction

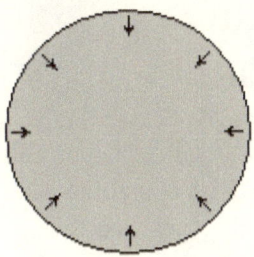

A spherical shape is the result of gravitational attraction. Gravity, or contracting space, will bring all mass to its lowest energy level. Everything will come to rest. The resulting shape of the object that will permit this effect, and pack mass most effectively, is a sphere. Any mass compacted into the smallest space possible ends up as a sphere. Consider that a result of gravity or contracting space - same thing.

Rotating objects take a slightly different form due to centrifugal forces. For example, the Earth has an equatorial bulge due to the rotation on its axis. The diameter of the Earth across the equator is 46 kilometers more than the diameter across the Earth's poles. This would give the Earth a slightly oblate shape as shown in Figure 29. The shape of this diagram is highly exaggerated in order to visualize the effect of the centrifugal force.

Figure 29 – Shaped by Centrifugal Force

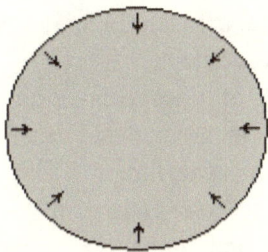

The equatorial bulge has nothing to do with gravity or tides, it's simply the result of centrifugal forces due to the rotation of the Earth on its axis.

This gives the Earth an oblate shape. Tidal bulges are superimposed on top of the equatorial bulge.

The Common View of Tides

Let's take a brief look at some of the most common explanations on the cause of tides. Some articles claim that tides result from a combination of gravitational and inertial forces due to the rotation of the Earth. The high tide nearest the moon is supposedly caused by the moon's gravitational pull, while the tide on the opposite side is caused by the inertia of the water as it's carried around the Earth. Some say that if the Earth were not rotating, there would be only one tide - on the side of the Earth facing the moon.

This particular theory does not hold up to technical scrutiny. See the article entitled "Tidal Misconceptions"[23] – by Donald E. Simanek. This article explains that the rotation of the Earth on its axis causes the "equatorial bulge." This is totally different from the "tidal bulge" induced by the Sun and moon. The equatorial bulge causes the oblate shape of the globe and is the same all around the equator. The "tidal bulge" is superimposed on top of the equatorial bulge, on opposite sides of the Earth.

Other articles offer an explanation that suggests there are centrifugal forces on the Earth resulting from its theoretical rotation about the barycenter. The barycenter is the center of gravity of the Earth and moon combined; and it is located about ¾ from the center of the Earth. It takes about 27.3 days to make a complete cycle. Because of the longevity of this cycle, the effect of its centrifugal force on tides would appear to be negligible. Let's call it non-existent.

A true understanding of tides cannot be explained by centrifugal forces. This is explained in great detail by Paolo Sirtoli in his paper entitled, "Tides and centrifugal force."[24]   "... We may rule out the centrifugal force as necessary to explain tides." – Paolo Sirtoli

The strongest theory and probably the most common is this: The cause of the tides is the difference in the gravitational force across the Earth. This theory acknowledges that gravity itself does not cause the tides directly. The moon's gravitational pull is strongest at the Earth's surface nearest the moon. Therefore the water there on the Earth's surface tends to bulge toward the moon. The gravitational pull is stronger at the center of the Earth than at the far side. So, the Earth tends to pull away from the water on the far side of the Earth, leaving the water to bulge on the far side. This theory will require some further explanation in the pages to follow.

These brief descriptions of the theories above cannot do justice to the extent as they were originally written. The thought here was just to review the popular beliefs currently held. We need not deal with the first two theories mentioned since they have already been addressed quite well in the articles mentioned above.

Gravitational Effects on the Earth

Let's examine the gravitational effect that the Sun and moon have on the Earth. The masses of the moon and Sun are 7.348×10^{22} and 1.989×10^{30} kilograms, respectively. The average distance from the Earth to the moon is 3.844×10^8 meters. The average distance from the Earth to the Sun is 1.496×10^{11} meters. Now, let's use the acceleration equation to calculate the gravitational accelerations on the Earth that are caused by the Sun and moon.

$$a = \frac{Gm}{d^2}$$

Moon	Sun
$\dfrac{6.6742 \times 10^{-11} * 7.348 \times 10^{22}}{(3.844 \times 10^8)^2}$	$\dfrac{6.6742 \times 10^{-11} * 1.989 \times 10^{30}}{(1.496 \times 10^{11})^2}$
$a_m = 3.3190 \times 10^{-5}$	$a_s = 5.9316 \times 10^{-3}$

a_m is the acceleration of the Earth caused by the moon's gravity.
a_s is the acceleration of the Earth caused by the Sun's gravity.

The Fundamental Force

The gravitational effect of the moon on the Earth is that the Earth accelerates toward the moon about .033 millimeters every second. The gravitational effect of the Sun on the Earth is that the Earth (and the moon) accelerates toward the Sun 5.9316×10^{-3} meters (about 6 millimeters) every second. That's about ¼ inch.

Remember that acceleration means a change in direction. So it doesn't mean that the Earth and moon are getting any closer to each other. It simply means that they maintain an orbit around each other without getting any closer.

Taking the ratio of gravitational accelerations of the Sun and moon we get:

$$a_s \; 5.9316 \times 10^{-3} \div a_m \; 3.3190 \times 10^{-5} = 178.7$$

This shows that the gravitational acceleration of the Earth caused by the Sun is 178.7 times greater than that caused by the moon. Therefore, we could postulate that if gravity was the cause of tides, then tides, caused by the Sun should be 178.7 times greater than those caused by the moon. Obviously this is not so. Therefore, the gravitational force does not directly cause the tides. There must be some other kind of relationship between gravity and tides.

The moon is so much closer than the Sun, so don't we have to factor that in? But we already have; look at the acceleration equation. It already contains the distance between the objects. The strength of gravity is inversely proportional to their distance squared. The distance is in the equation and already accounted for. Let's continue.

Comparing Distances

Let's take a look at the distances.

Sun	Distance from the Earth	=	1.496×10^{11}
Moon	Distance from the Earth	=	3.844×10^{8}
	Ratio		389.18

The moon is 389 times closer to the Earth than the Sun. Some of the common explanations of tides will divide the two ratios to get the proportional strength of the tidal force. Let's do that.

$$\frac{\text{Ratio of gravitational strength}}{\text{Ratio of distance from the Earth}} = \frac{178.7}{389.18}$$

$$\text{Ratio} \qquad\qquad 0.459$$

So there it is - the ratio of gravitational strengths divided by the ratio of distances is 0.459. The implication here is that the tides on the Earth caused by the Sun are only 45.9% as high as those caused by the moon. This apparently is quite accurate based on actual measurements. But the reasoning is a little difficult to digest; so let's do this calculation in a different way that may give us a greater understanding. Let's divide the masses of the Sun and moon by their distance from the Earth cubed. This is basically the same operation that was just done above, but with the gravitational constant factored out.

<table>
<tr><td align="center"><u>Moon</u></td><td align="center"><u>Sun</u></td></tr>
<tr><td align="center">$\dfrac{7.348 \times 10^{22}}{(3.844 \times 10^{8})^{3}}$</td><td align="center">$\dfrac{1.989 \times 10^{30}}{(1.496 \times 10^{11})^{3}}$</td></tr>
<tr><td align="center">$d_m = 1.2937 \times 10^{-3}$</td><td align="center">$d_s = 5.9407 \times 10^{-4}$</td></tr>
</table>

d_m is the density of space around the moon to the Earth.
d_s is the density of space around the Sun to the Earth.

Notice that the answers are denoted by d_m and d_s. We divided mass by volume, so these numbers are densities. Not the density of the moon or Sun, but the density of space surrounding the Sun or moon with the radius being the distance to the Earth.

Taking the ratio of these two space densities will give the same result:

d_s $5.9407 \times 10^{-4} \div d_m$ $1.2937 \times 10^{-3} = 0.459$

This is the same ratio as previously obtained. However, the one derived in this case can be viewed as resulting from the density of space near the Earth, not gravitational strength. Notice that even the gravitational constant has been factored out of the calculation. This suggests that the cause of tides could somehow be related to the geometry of space.

So, the force that causes tides, (the force of tides) appears for some reason to be different than the force of gravity. The force of gravity is inversely proportional to the distance *squared*. The force of tides is inversely proportional to the distance *cubed*. This is still a matter of proportionality, but it does not explain why we have high tides on opposite sides of the Earth.

Gravitational Difference Across the Earth

Many articles on tides attempt to explain high tides on opposite sides of the Earth by the gravitational differential across the diameter of the Earth. In other words, the gravitational force of the moon on the side of the Earth nearest the moon is stronger than the moon's gravitational force on the far side of the Earth. That difference, they say, is the cause of tides on both sides of the Earth. This difference is often referred to as the gravitational gradient across the Earth, or the tidal force, or force of tides.

The gradient of the gravitational force across the diameter of the Earth is extremely small, but can be easily calculated. We will do that shortly. But first we should understand the reasoning used to justify this theory. So now, let's discuss the argument used to promote this theory.

The proponents of this theory would say the cause of tides is the difference in the gravitational force from one side of the Earth to the other. The gravitational pull is stronger at the Earth's surface nearest the moon than at the Earth's center. Therefore the water on the surface gets pulled away from the Earth and tends to bulge on the side of the Earth facing the moon.

I would acknowledge that there is in fact a very small difference in the gravitational force across the Earth. But is that difference enough to cause the tides? We shall see. Let's calculate the difference and quantify the height of the bulge. Each particle of mass on the surface of the Earth nearest the moon, no matter what size, even a single drop of water, will accelerate toward the moon at the rate specified by Newton's equation.

We have already made the calculation for the gravitational acceleration of the center of the Earth toward the moon. It is $3.3190\text{x}10^{-5}$, or approximately .033 millimeters. Now let's do the same calculation using the equation below to calculate the acceleration of particles at the surface of the Earth nearest the moon, then again on the opposite side. We'll assume the distance between the centers of the Earth and moon is $3.844\text{x}10^8$ meters. We can add or subtract the radius of the Earth to get the distance we need for the near and far side of the Earth. The radius of the earth is $6.371\text{x}10^6$ meters. Once again we will use the acceleration equation to calculate the accelerations.

$$a = \frac{Gm}{d^2}$$

<u>near side of moon</u>   <u>far side of moon</u>

$$\frac{6.6742\text{x}10^{-11} * 7.348\text{x}10^{22}}{(3.844\text{x}10^8 - 6.371\text{x}10^6)^2}$$ $$\frac{6.6742\text{x}10^{-11} * 7.348\text{x}10^{22}}{(3.844\text{x}10^8 + 6.371\text{x}10^6)^2}$$

$$a_n = 3.4317\text{x}10^{-5}$$ $$a_f = 3.2116\text{x}10^{-5}$$

a_n is the acceleration on the side of the Earth nearest the moon.
a_f is the acceleration on the side of the Earth farthest from the moon.

The acceleration of the side of the Earth nearest to the moon is about .034 millimeters, while the acceleration at the center of the Earth is approximately .033 millimeters. The difference between the two: .034 - .033, is about .001 millimeters.

Notice that the gravitational acceleration across the Earth decreases as we get farther away from the moon. Since the acceleration is less at the center of the Earth than at the near side, there is no tidal force, or differential gradient, to cause a bulge at the surface of the Earth on the near side. In other words, there is no "push" from within the Earth to cause a bulge. The only force being applied to cause a bulge is the gravitational force. The height of the bulge is calculated by Newton's law of gravity to be .001 millimeters. So, the maximum high tide caused by gravity could only be about .001 millimeters; or about 1/1000th of a millimeter. That's it – no more.

Let's continue with the same argument for the far side. The proponents would say that the gravitational pull is stronger at the center of the Earth than at the far side. So the Earth tends to pull away from the water on the far side of the Earth, leaving the water to bulge on the far side

As we have seen from Newton's equation for gravity, the center of the Earth accelerates toward the moon .033 millimeter per second – no more. If the Earth accelerated more than this, there would have been a very massive collision between the Earth and the moon a long time ago. Since the Earth and moon are still here, we can assume that has not happened yet, and that the laws of Sir Isaac Newton are still valid.

From the calculation above we see the acceleration of the far side of the Earth toward the moon is about .032 millimeters. The difference in gravitational acceleration between the center of the Earth and the far side is .033 - .032 = .001 millimeters; also about 1/1000th of a millimeter. So the maximum high tide caused by gravity on the far side of the Earth could only be about 1/1000th of a millimeter. That's it – no more. Also, since the gravitational acceleration of the Earth toward the moon is greater at the center of the Earth than at the far side, there is no tidal force, or push from the Earth, to cause a bulge at the surface of the Earth on the far side.

Remember how gravity works, whether we consider it a force or an effect of contracted space, it still works the same. All the particles of mass on the Earth, every single atom, accelerate together as a whole toward the moon about .033 millimeters every second. The .002 millimeter difference across the diameter of the Earth is too small to be the cause of tides. Therefore, this explanation of the cause of tides cannot possibly be valid.

If you are confused right now – you should be. The line of reasoning we have just gone through shows that the gravitational difference across the Earth is insignificant – too small to cause tides. In fact, none of the explanations for the cause of tides are convincing. However, the reality is that there are tidal bulges – so there must be a reason. And there is, but the reason is not gravity. Bear with me just a little bit longer.

Why Should There Be Tides?

So far we have not seen a reasonable explanation for the cause of tides. Before going further, let's ask the following question. Why is there any tide at all? Let us consider the Earth's gravitational 'force' on its own bodies of water. It is so much stronger at the surface of the Earth, than either the moons or the Suns gravitational pull. The Earth's gravitational strength is more than sufficient to overcome the gravitational influence of both the Sun and the moon and keep everything in a spherical or oblate shape.

We can measure the strength of gravity at the surface of the Earth by applying the acceleration equation again. The mass of the Earth is 5.974×10^{24} kilograms and the Earth's radius is 6.371×10^{6} meters.

$$a = \frac{Gm}{d^2} = \frac{6.6742 \times 10^{-11} * 5.974 \times 10^{24}}{(6.371 \times 10^{6})^2}$$

$a = 9.8231$ Meters per second per second.

This is the acceleration caused by the Earth's gravity on bodies at the Earth's surface. It can also be considered the strength of the Earth's gravity at the Earth's surface. Compare this with the gravitational effect caused by the moon at the Earth's surface. The value for the acceleration caused by the moon on the near side of the Earth is taken from the calculated value we obtained in the previous section. It is the acceleration on the side of the Earth nearest the moon. Let's take the ratio of both accelerations:

Acceleration caused by Earth	=	9.8231
Acceleration caused by moon	=	3.4317×10^{-5}
Ratio		2.8624×10^{5}

This shows the difference in gravitational strengths at the surface of the Earth on the side nearest to the moon. If we view these as opposing forces, we can see the Earth's gravitational acceleration is 286,000 times stronger than the moon's acceleration. There is simply insufficient acceleration toward the moon to overcome the Earth's gravitational attraction and cause a tide. If we consider the accelerations as opposing vectored forces we can do some simple arithmetic to prove this point. Subtract the acceleration caused by the moon from the Earth's acceleration at the surface of the Earth: $9.8231 - 3.4317 \times 10^{-5} = 9.8230$. The difference would appear to be too small to be of any consequence whatsoever. Its effect would be too small to measure, and any effect would go unnoticed.

How can we reasonably say that even one drop of water in the vast oceans of the Earth will move just one millimeter closer to the moon? Why should it, when there is a much stronger force attracting every drop of water in the opposite direction? There seems to be no justification whatsoever in assuming there would be any measurable movement of any kind toward the moon. It would seem that we can state quite clearly, and with a high degree of certainty, that it is not the gravitational fields of the Sun or moon that cause the tides. There must be something else – but what?

Physicists will tell us when doing calculations using vectored forces, in order to arrive at a correct conclusion; all forces must be taken into account. So, let's do that. But first let's eliminate those forces that cannot account for the tides. Let's eliminate centripetal and centrifugal forces; then let's eliminate inertial forces, and then finally let's eliminate gravity. We're left with ... nothing. But there has to be *something* causing the tides. Yes, of course there is; but this missing something has never been included in the equation for tides until now.

Chapter 15

The Cause of Tides

Why tides on both sides?

Seeing Both Sides

Since the days of Newton we have explained away the tides as being caused by gravity; we had to - what else was there? But in the previous chapter we learned that gravity is *not* the cause of tides. The gravitational fields of the Sun and moon certainly have an effect, but they cannot cause tides. But there is a missing piece to this puzzle that has never been considered before. The missing piece to the equation is the expansion of space. Is that possible, and if so how can expanding space cause tides? In this chapter we will explore the true cause of tides and come to understand what parts space expansion and space contraction play in the distortion of space, and the distortion of the Earth, to bring about tides.

How could we have overlooked expanding space? In 1917, Dutch cosmologist Willem De Sitter saw a peculiar phenomenon in Einstein's equations that *implied* an expanding universe; But Russian mathematician Alexander Friedmann said no, Einstein's theory *demands* an expanding universe. Einstein could not accept their analysis until 1925 when Edwin Hubble gave us the first clue. He announced his findings about galaxies receding from us; this indicated an expanding universe. More recent observations in 1989, by Saul Perlmutter and his team of astronomers confirmed the expansion. They also found that the universe is expanding at an accelerating rate.

Space is expanding and so the search began for "dark energy;" that mysterious force behind the expanding universe. The clues and evidence were there, and have been there for some time. Einstein pointed the way, but even he himself didn't fully realize all the implications of relativity; the theory was too new. We needed time to grasp the reality and implications of this most remarkable discovery. We needed time to understand, analyze, and digest it. With all that done, no one saw a need to apply our newly acquired knowledge to explain tides. Perhaps we had become too comfortable and satisfied with existing theories and explanations; and we have not applied our new knowledge where needed.

Now we know that space is expanding; but what we never knew before is how expanding space effects matter. Without this understanding space expansion has been virtually ignored in all our physical scientific equations. What we still may not realize is that we have some expanding space in our own back yard. The effect of expanding space is here, and has been here all along – we call them tides. We just didn't know that tides are an effect of expanding space.

Alone in Space

Space is expanding all around us and we never knew it. Why not? – Probably because it's expanding in all directions at the same rate. So we have a situation around planets as depicted in Figure 30 – Expansion and Contraction. This figure shows the expansion and contraction of space in opposite directions. We have a large mass which is contracting space toward itself. We call this gravity. Then we have expanding space. From any object's point of view space is expanding away from the object. This, of course, would apply to all objects, and all mass, everywhere.

Figure 30 – Expansion and Contraction

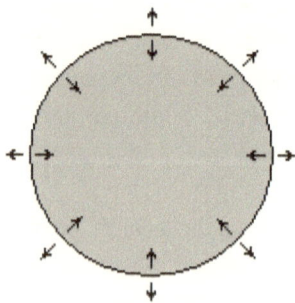

Figure 30 shows that these two characteristics of space, expansion and contraction, are in diametrically opposite directions. What we must understand now is that we have two opposing spatial changes; one inward (toward mass), the other outward (away from mass). How do they affect the body of mass?

In space, around any massive object remotely located from other objects, the only observable evidence of contraction or expansion of space is gravity. The rate of space contraction would be the same all around the planet. The rate of spatial expansion would also be the same all around the planet. A balance is achieved that would go undetected. Since the rate of contraction is much greater than the rate of expansion, the only effect one would see, or feel, is gravity. However, when two bodies in space come into close proximity, an imbalance is introduced; the effects become noticeable. We call these effects "tides."

Crude Balloon Experiment

The question still remains: how does expanding space cause the tides. First let's demonstrate this with a simple balloon experiment. Take a 12 inch party balloon and glue two round disks to it with a little space between them. Inscribe a circle on the balloon around each disk. Watch the effect as the balloon is blown up.

Figure 31 – Expanding Balloon

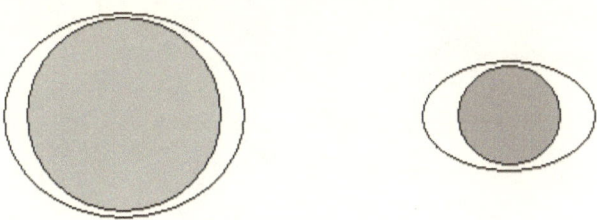

As the balloon is blown up, envision the expanding balloon as expanding space. Notice that the tension of the balloon between the two attached disks is greater than the tension perpendicular to the two disks. The two disks will remain closer to each other than the expansion of the balloon on the opposite side. It would appear the two disks are giving resistance to the expansion. Notice the circles inscribed on the balloon around the disks. They are expanding in an elliptical pattern, not a circular one. Also notice the elongation of the ellipse occurs on both sides of the disks along the axis connecting the disks.

This is a rather crude experiment, but we're trying to illustrate a three dimensional effect on a curved two dimensional surface. But, let's analyze what happened. First of all we can see and feel that the tension on the balloon is greater on the axis of the two disks. This causes the elliptical shape taken by the expanding circles. The reason for the increased tension is because the two disks are inhibiting the balloons expansion in those areas. All the growth that occurs around the circumference through the axis of the disks must come from the balloon not covered by the disks. There's less balloon surface that can be used for expansion than on the circumference perpendicular to the plane. That explains the elliptical shape around the two glued disks.

What we're seeing in Figure 31 is only the expansion of space. It does not show the effect of space contraction. Contracting space would try to keep the mass bound in the smallest space possible. Expanding space would try to expand the space on the axis between the two centers of the objects. The following figure shows both, the expansion of space and the contraction of space.

Figure 32 – Expanding and Contracting Space

In this case we can see the shape the bodies in space will take when they come into close proximity. The shape goes from spherical to ellipsoidal. The closer the bodies get, the more pronounced the ellipsoidal shape. The volume of each object remains the same, but the shape of the volume changes. Of course Figure 32 is only an illustration from an expanding balloon that shows how physical objects would be reshaped. The actual expansion of the Earth at the location of the high tides is very slight and measured in centimeters.

Two objects in close proximity in space cause an imbalance in the structure of space along the axis connecting their centers. As shown in Figure 32, we can see the amount of expansion, along the axis of the two objects is greater than that of the line perpendicular to the axis. This simple experiment shows the restructuring of space in the vicinity of objects in close proximity. As space is restructured, so are the objects in space.

The Structure of Space

As Einstein said, the properties of space are determined by matter. In different words, the form or shape of space is dictated by the matter it contains. There are two components at work restructuring space, and the objects in it. These two components working in opposite directions, are space contraction and space expansion. As has already been described, the contraction of space around massive objects tends to keep them in the most compact size possible. Space expansion distorts, or restructures, the space around massive objects in close proximity which causes the tides. But we also need to realize that the restructuring of space and objects is not a forceful occurrence. Space simply conforms in shape to the matter contained within it. And matter also conforms to space, just as space conforms to matter.

Consider the finer things in life – perhaps a cold bottle of your favorite beer. Pour the beer into a frozen glass. Ah – life is good! How much force was spent by the beer in reconfiguring itself into the new shape of the glass? Not much – not even worth considering. The beer just filled its new container, no matter what the shape – no energy or effort required. So it is with the Earth and all other large objects in space. Matter simply fills the space provided while assuming the most compact size possible. In this case the most compact size possible is an ellipsoid. Space may alter its shape, but there is no pulling, no pushing, and no energy expended in forcing matter to conform in shape. No massive forces need be applied. Matter just flows, or falls, to the lowest energy level, to be at rest.

So here we have the cause of tides. Space is expanding in all directions, but the space occupied by objects in close proximity expands more on the axis of the two objects than on the perpendicular to the axis. This imbalance causes a reconfiguration of the space containing these objects. The resulting space of the object is slightly warped into an elliptical shape with the major axis being the line between the centers of the two objects causing the spatial distortion. The mass of an object conforms to the space provided and fills the space to its lowest energy level. When all mass is at rest, the resulting shape of the object will be slightly ellipsoidal. That is the cause of tides.

Tidal Thoughts

In the chapter on Misunderstanding Tides we examined the rational explanations for the cause of tides. There were none that could withstand critical scrutiny. We examined the most common belief and applied the mathematics. We could see no justification for tides.

In this chapter a different cause has been presented. This explanation is based on Einstein's notion that the geometrical properties of space are determined by matter. We saw that the gravitational effect between bodies in space is not sufficient to cause tides, but that the combination of space contraction and space expansion will distort the space around bodies in close proximity.

This restructuring of space is not a forced affair. Empty space does not force mass into any particular shape. There is no pulling or pushing. The shape space will take is based on space contraction and expansion. The mass contained in space conforms to the shape of that space.

Chapter 16

The Amplitude of Tides

Good tidings to all.

The Theoretical Amplitude

Current thinking suggests that the tides are caused by the differences in the gravitational attraction across the Earth. Scientists use this model to calculate a theoretical value for the amplitude of the tide. This is assuming an Earth that is not rotating, covered with water, and with all land masses deeply submerged. Here's an equation given by Eugene I. Butikov in his paper entitled, "A dynamical picture of the oceanic tides".[25] This equation gives us the amplitude of the high tide on the Earth. We'll apply this equation for both the Sun and the moon.

Equation 28 – Amplitude of Earth's Tide

$$a = \frac{3}{2} * \frac{\text{mass of object}}{\text{mass of Earth}} * \frac{(\text{radius of Earth})^3}{(\text{distance from Earth})^3} * \text{radius of the Earth}$$

a   is the amplitude of the tide.

I took the liberty of restructuring the equation to make a point. The restructured equation is listed below; and below that the calculations are made for the moon and the Sun. The reason for restructuring the equation is to show that the mass of an object divided by its distance from the Earth *cubed* is part of this equation. This form of the equation also shows a ratio of the density of space surrounding the object and the density of the Earth.

Equation 29 – Amplitude of Earth's Tide

$$a = \frac{3}{2} * \frac{\dfrac{\text{mass of object}}{(\text{distance from Earth})^3}}{\dfrac{\text{mass of Earth}}{(\text{radius of Earth})^3}} * \text{radius of the Earth}$$

    a   is the amplitude of the tide.

We often speak of the "tidal force" as the cause of tides. Take a close look at the equation above. There is no force of any kind specified in the equation. This indicates that the tides are the result of the structure of space. That is why I prefer the term "tidal effect" instead of "tidal force." The tidal effect is caused by the geometry of space. The structure of space determines the shape of objects in space. Objects take on an ellipsoidal shape when in close proximity to other massive objects; the closer the object the more pronounced the ellipsoidal shape; the greater the mass of the object the greater the ellipsoidal shape.

Let's value the equation to get the amplitude of the high tide on the Earth caused by the moon. The only values we need are the masses of the two objects, their distance apart, and the radius of the object whose tidal amplitude is being obtained.

These are the values required for the equation:
- Mass of the Earth 5.970×10^{24} kilograms
- Mass of the moon 7.348×10^{22} kilograms
- Distance apart 3.844×10^{8} meters
- Radius of the Earth 6.371×10^{6} meters

$$a = \frac{3}{2} * \frac{\dfrac{7.348 \times 10^{22}}{(3.844 \times 10^{8})^3}}{\dfrac{5.974 \times 10^{24}}{(6.371 \times 10^{6})^3}} * 6.371 \times 10^{6}$$

$$a = 53.51 \text{ centimeters}$$

This calculation shows that the theoretical amplitude of the high tide from the moon's tidal effect is about 53.5 centimeters.

We'll now use the same equation for the Sun. Here we are replacing the moon's mass and distance from the Earth, with the Sun's mass and its distance from the Earth. Everything else remains the same.

These are the values required for the equation to calculate the high tide on the Earth caused by the Sun:

- Mass of the Earth 5.970×10^{24} kilograms
- Mass of the Sun 1.989×10^{30} kilograms
- Distance apart 1.496×10^{11} meters
- Radius of the Earth 6.371×10^{6} meters

$$a = \frac{3}{2} * \frac{\dfrac{1.989 \times 10^{30}}{(1.496 \times 10^{11})^3}}{\dfrac{5.974 \times 10^{24}}{(6.371 \times 10^{6})^3}} * 6.371 \times 10^{6}$$

$$a = 24.57 \text{ centimeters}$$

This calculation shows that the theoretical amplitude of the high tide from the Sun is about 24.5 centimeters. Taking the ratio of the two will give us the same results, $(24.575 \div 53.510 = 0.459)$, we've seen before. This is the same ratio we obtained when calculating the ratio of space densities in the section on Comparing Distances, in the chapter on Misunderstanding Tides.

In the calculation for the Sun, the only values that changed in the equation were the mass of the Sun and its distance from the Earth. The mass of an object and its distance from the Earth are part of the equation that determines the amplitude of the tide. This can also be restated as: the density of space near the Earth and the density of the Earth determine the size of the tides. The amplitude of the tides on the Earth are directly related to the mass of an object and inversely related to the cube of the object's distance from the Earth.

We have just calculated the amplitude of the high tides on the Earth caused by the Sun and moon. When the Sun and moon are aligned with the Earth the amplitude of the tides are cumulative. These unusually high tides occur twice a month during the new and full moons; they are called the Spring Tides. When the Sun and moon are at right angles with the Earth at the focus, the tides are unusually low. Their tidal effects minimize each other and these tides are called Neap Tides.

The Measured Amplitude

It is an extremely difficult task to accurately measure the amplitude of the high tides caused by the Sun and moon. The best location to take measurements is in the middle of the ocean. There are many factors that must be accounted for: latitude of the location where the measurement is being made, the depth of the ocean, ocean currents, the topology of the ocean floor, centrifugal force, etc. Obtaining an accurate measurement of the tides is similar to the difficulty in measuring the gravitational constant.

An article in infoplease encyclopedia entitled; The Magnitude and Effects of Tidal Ranges[26] informs us that the typical tidal range in the open ocean is 2 feet.

2 feet is about 61 cm.

An article on Moon Tides - How the Moon Affects Ocean Tides[27]... states; "Offshore, in the deep ocean, the difference in tides is usually less than 1.6 feet."

1.6 feet is about 49 cm.

In a previously quoted article, Tidal Misconceptions[2] Donald E. Simanek states that the amplitude of tides in deep mid-ocean is about 1 meter.

One meter is 100 cm. The theoretical high tide from both the Sun and moon combined would be about 53.51 + 24.57 = 78 centimeters.

Considering the difficulty of measurement, it would seem the theoretically calculated value of 53.5 cm for the tidal amplitude caused by the moon is fairly close to the measured values.

The Amplitude of the Lunar Tide

We are about to calculate the amplitude of the high tide on the moon caused by the Earth. This may seem quite odd at first because the moon has no water on its surface, and we normally associate the tides with the flow of water. The flow of water is what we can observe and naturally relate with tides. But the flow of water is only part of the story.

In the previous chapter we described the structure of space as the cause of the restructuring of objects in space. When we consider two objects in close proximity, like the Earth and moon, we have to realize that both those objects will be slightly restructured. Both objects become slightly ellipsoidal in shape. That restructuring creates the tide. The restructuring affects the whole object, not just the water. On the Earth during the high tide, whole continents will rise as much as 30 centimeters, about one foot. The shape of the whole Earth changes, and the shape of the moon is changed, even more so.

This constant re-sculpturing of the Earth often results in cracks in the Earth's surface. Occasionally these cracks will cause tremors and earthquakes. When the cracks are deep enough then we can have volcanic activity.

Up until now we have been discussing the amplitude of the tides on the Earth. Let's now use Equation 28 to calculate the amplitude of the tides on the moon due to the proximity of the Earth and the Sun. First we'll make a slight adjustment to the equation by combining the radius of the object under measurement; so now the radius is raised to the 4th power instead of being cubed. This does not aid in our understanding, just simplifies the calculations.

$$a = \frac{3}{2} * \frac{\text{mass of close object}}{\text{mass of tidal object}} * \frac{(\text{radius of tidal object})^4}{(\text{distance between})^3}$$

The mass of the moon is 7.348×10^{22} kilograms; and its radius is 1.738×10^6 meters.

The amplitude of the tide on the moon caused by the Earth:

$$a = \frac{3}{2} * \frac{5.974 \times 10^{24}}{7.348 \times 10^{22}} * \frac{(1.738 \times 10^6)^4}{(3.844 \times 10^8)^3}$$

a = 19.59 meters

The high tide on the moon is almost 20 meters, or about 64 feet. This would be extremely hard to measure accurately considering the moon does not have a smooth surface. We don't see any changes of the moon's dimensions because the moon is tidally locked into the Earth. This means that the moon's tidal bulge is always facing the Earth, so no tidal changes are taking place across the surface of the moon. The ellipsoidal shape taken on by moons close to a planet is the reason why they are tidally locked in. Satellites with ellipsoidal shapes are in stable positions when their semi-major axis is in alignment with the parent planet.

Now let's look at the amplitude of the high tide on the moon caused by the Sun. The mass of the Sun is 1.989×10^{30} kilograms and its distance from the moon is 1.496×10^{11} meters.

$$a = \frac{3}{2} * \frac{1.989 \times 10^{30}}{7.348 \times 10^{22}} * \frac{(1.738 \times 10^6)^4}{(1.496 \times 10^{11})^3}$$

a = 11 centimeters

The Sun's tidal effect on the moon is insignificant compared to the Earth's because it's so much farther away from the moon than the Earth. Remember, the tidal effect diminishes rapidly with distance, because of the inverse of the distance *cubed* relationship.

The Amplitude of Tides on Io

Io is a satellite of Jupiter. This is an interesting object to study because it is a little larger in size compared to our moon and its distance from Jupiter is just a little greater than the distance our moon is from the Earth. The point of interest is Io's close proximity to a very massive planet. Io is pictured in Figure 33.

Figure 33 – Io

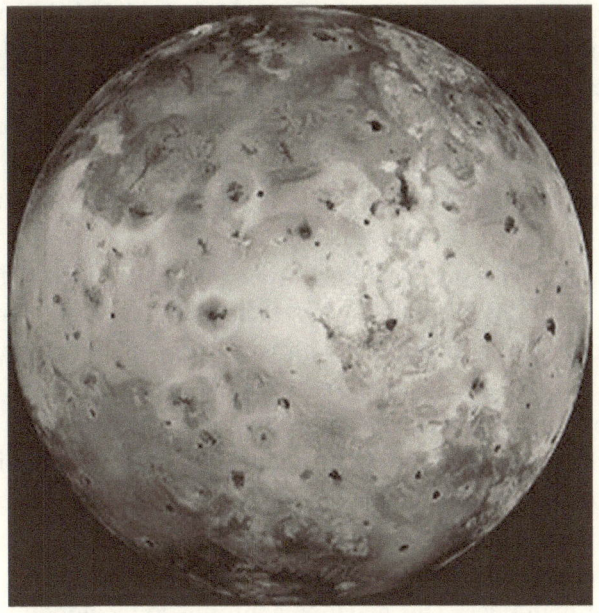

Image Credit: NASA

http://nssdc.gsfc.nasa.gov/imgcat/hires/gal_p50559.jpg

Let's calculate the high tide on Io caused by the massive planet Jupiter. These are the values required for this calculation:

- Mass of Jupiter 1.899×10^{27} kilograms
- Mass of Io 8.932×10^{22} kilograms
- Distance apart 4.216×10^{8} meters
- Radius of Io 1.821×10^{6} meters

$$a = \frac{3}{2} * \frac{1.899 \times 10^{27}}{8.932 \times 10^{22}} * \frac{(1.821 \times 10^{6})^{4}}{(4.216 \times 10^{8})^{3}}$$

$a = 4679$ meters

The tidal bulge of Io is about 4.6 kilometers. But that is still relatively small compared to the radius of Io, which is 1,821 kilometers. The amplitude of the moon's tidal bulge caused by the Earth is less than 20 meters. The amplitude of the tide on Io is considerably larger than that of our moon. The main difference in tidal size can be attributed to the differences in masses of the Earth and Jupiter. Jupiter is 317 times more massive than the Earth, and Io's tide is about 240 times higher than the moon's.

Both Io and its sister satellite Europa are tidally locked into Jupiter. This means the tidal bulge is always facing Jupiter. However, these two moons have frequent encounters with each other and neighboring moons Ganymede and Callisto. These close encounters cause friction as the tidal bulge makes minor shifts in position. Internal heating occurs in Europa, which is believed to keep the water under the surface in a liquid or slushy state. Io is closer to Jupiter than Europa, so the internal heating of Io is much more dramatic, causing volcanoes and lava flow.

In this chapter we analyzed an equation that is used to calculate the amplitude of high tides. This equation conforms to the notion that the tides are the result of the structure of space. In the equation we saw a ratio of the density of space surrounding an object in close proximity and the density of the Earth. The densities of the space determines the structure of the space; the structure of the space determines the shape of objects in that space. When objects come in close proximity they take on an ellipsoidal shape; that ellipsoidal shape is the tide.

Chapter 17

The Law of Tides

The tide has turned!

Great Expectations

Before proceeding with this chapter there is something I need to briefly reiterate. My intention is not merely to reveal how and why certain scientific laws work the way they do, but also to show how I came to the realization of those conclusions. The discovery process is sometimes very slow, just one step at a time, and sometimes those steps are very small.

I had been thinking about the cause of tides for many years before ever hearing about the Pioneer anomaly or the galaxy rotation problem. I could not accept the existing explanations for the cause of tides, so I conceived my own theory based on faith, without proof. I held to this belief for years with no evidence whatsoever. This made me realize how much science is based on faith. Einstein must have had some faith in his theory of relativity before he could provide evidence to support his case. Scientists in search of dark matter must believe it's out there, otherwise they wouldn't waste their time searching for it. They have great expectations. So did I.

In the chapter Misunderstanding Tides I showed through mathematical analysis, that gravity cannot by itself cause the tides. Then in the chapter The Cause of Tides I claimed that tides are caused by the expansion of space based on a simple balloon experiment. That balloon experiment may be an acceptable illustration of how the tides work, but is hardly considered evidence. So far there has not been enough evidence presented to prove the case for the expansion of space causing the tides.

I thought someday someone would find the necessary evidence to reveal the real culprit causing the tides. As hard as I tried I could not find it myself. But sometimes clues turn up in unexpected places, and in unexpected ways. So it is here. The needed evidence came not from studying the dynamics of the tides, but from the investigation of the galaxy rotation problem. The equation for space expansion within the solar system is given in the last section of the chapter Space Expansion.

So now, armed with an equation for space expansion, the necessary evidence to convict the real culprit causing the tides is at hand. We can now proceed with this chapter and lay down the law on tides.

The Tide Turns

Over the course of the last few chapters we have learned that gravity is not the cause of tides. I presented instead the hypothesis that tides are caused by the expansion of space. In those chapters only thoughts and ideas were offered suggesting that the tides are the result of the structure of space, and the structure of space is the result of gravity and expanding space; there was no evidence to make a convincing case. Now the tide has turned.

We now have a way of calculating the acceleration from expanding space. But first let's go back to Butikov's equation in The Amplitude of Tides chapter that showed us how to calculate the high tides; we will use that equation as a spring board to help in our understanding. Here again is the equation given by Eugene I. Butikov in his paper entitled "A dynamical picture of the oceanic tides."[28] This equation applies to any objects in close proximity in space. We've already applied it to both the Sun and the moon in determining the high tide on the Earth from each.

Equation 30 – Amplitude of Earth's Tides

$$a_t = \frac{3}{2} * \frac{\text{mass of object}}{\text{mass of Earth}} * \frac{(\text{radius of Earth})^3}{(\text{distance from Earth})^3} * \text{radius of Earth}$$

a_t is the amplitude of the high tide.

A formal description of this equation is not being used because we are about to make multiple revisions. It would be easier to follow the revisions when the equations are presented in a more informal way. Let's revise this equation again just as we have done before. The revised equation is from the chapter, "The Amplitude of Tides" and is shown in Equation 31.

Equation 31 – Amplitude of Earth's Tides

$$a_t = \frac{3}{2} * \frac{\dfrac{\text{mass of object}}{(\text{distance from Earth})^3}}{\dfrac{\text{mass of Earth}}{(\text{radius of Earth})^3}} * \text{radius of Earth}$$

a_t is the amplitude of the high tide.

In this case we are using the Earth and the moon as our objects of reference. Equation 31 shows us that the amplitude of tides on the Earth is a result of the structure of the Earth and the structure of space around the Earth. There is no force in this equation to cause the tides, but this equation does work quite nicely. Let's make another adjustment to show Equation 31 in terms of acceleration. If we multiply the mass of each object by the gravitational constant we will have the equation for acceleration embedded in both the numerator and denominator. We can then rewrite Equation 31 as follows:

Equation 32 – Amplitude of Earth's Tides

$$a_t = \frac{3}{2} * \frac{\dfrac{\text{acceleration of object}}{\text{distance from Earth}}}{\dfrac{\text{acceleration of Earth}}{\text{radius of Earth}}} * \text{radius of Earth}$$

a_t is the amplitude of the high tide.

We made this adjustment in order to think of tides in terms of acceleration. Let's think of acceleration as the strength of gravitational attraction. Now let's interpret this equation to see if we can explain the cause of tides. We divide the gravitational attraction of the moon on the Earth by the distance between them. Then we divide the gravitational attraction at the surface of the Earth by the radius of the Earth. This gives a ratio of two accelerations; both of which result from the contraction of space. Why we divide these accelerations by distance is still not clear.

In essence we have two gravitational attractions in opposite directions: the gravitational attraction of the moon "pulling" on the Earth, and the Earth's gravitational attraction at its surface, attracting objects toward its center. We then take the ratio of these two attractions and multiply that by the radius of the Earth and 3/2, to get the amplitude of the high tide on Earth. The problem remains; gravitational attraction cannot possibly explain the high tide on the opposite side of the Earth. Where the two attractions are in the same direction there is no logical explanation for a high tide on the opposite side of the Earth.

Figure 34 – Tides by Gravity, illustrates the problem. The flow of space contraction is toward the center of the objects of reference. The space flow at the surface of the earth, on the opposite side of the moon, is toward the moon, so there is no logical reason for a high tide at that location.

Figure 34 – Tides by Gravity

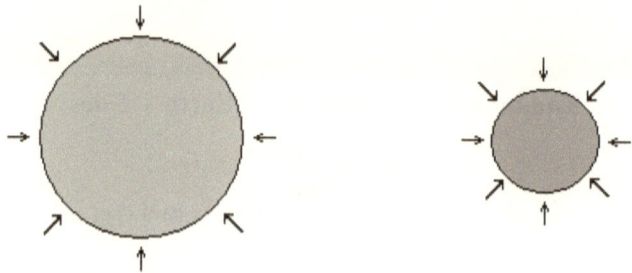

The Tidal Acceleration

In the previous section we have unsuccessfully attempted to explain the high tides on both sides of the Earth due to the contraction of space. Now we will explain why we have the high tides on both sides of the Earth because of the combination of space contraction and expansion. Let's take a look at the two equations used to calculate the accelerations from space expansion and space attraction. The following equation for acceleration caused by space expansion that was given in Equation 27 – Acceleration via Expanding Space.

$$a = \frac{GM}{pc_x * d}$$

Where:

a   is the acceleration due to expanding space.
G   is the Gravitational Constant
M   is the Mass from which the expansion emanates
d   is the distance between the object and the center of expansion
pc_x is the length of a short parsec – 2.9291×10^{16}

Below is Newton's equation for gravitational acceleration that we've been using all along.

$$a = \frac{GM}{d^2}$$

If we were calculating the accelerations of the same object using both the above equations, the products would be working off the same mass and gravitational constant. We can combine them to get what we can call the tidal acceleration by multiplying the denominators to get the following equation:

Equation 33 – Tidal Acceleration

$$a = \frac{GM}{pc_x * d^3}$$

This is not a multiplication or addition of equations. This is a new equation that gives us the acceleration caused by space expansion and space contraction. This is the tidal acceleration that diminishes in strength by the combined distances of space contraction and expansion, (i.e. pc_x * d^3). In essence we have combined the acceleration of gravity and the acceleration of expanding space from the same object into a single equation.

Let's utilize Equation 33 in the numerator and denominator of a new tidal equation which will give us the amplitude of the high tide based on tidal acceleration:

Equation 34 – Amplitude via Tidal Acceleration

$$a_t = \frac{3}{2} * \frac{\text{tidal acceleration from moon}}{\text{tidal acceleration from Earth}} * \text{radius of Earth}$$

We can view part of this equation as the ratio of tidal accelerations. In essence this ratio is the distortion of space. We apply that distortion across the radius of the Earth by multiplying by the Earth's radius. This gives us the amount of distortion on the surface of the Earth. Since the distortion is not evenly divided on the Earth's surface, we need to multiply by 3/2 to obtain the amplitude of the tide at its highest point; that is the amplitude of the high tide.

Now we have an equation that incorporates the accelerations from gravity and expanding space. This is the balance of space contraction and space expansion. Figure 35 – Tides via Space Expansion illustrates the balance of space flow. Contracting space keeps the Earth and moon in the most compact size. The expansion of space stretches the surrounding space, resulting in the restructuring of both the Earth and the moon, into ellipsoidal shapes.

The combined effects of space contraction and space expansion are shown in Figure 35 – Tides via Space Expansion. This is essentially the same effect that we saw in the simple balloon experiment that was demonstrated in a previous chapter.

Figure 35 – Tides via Space Expansion

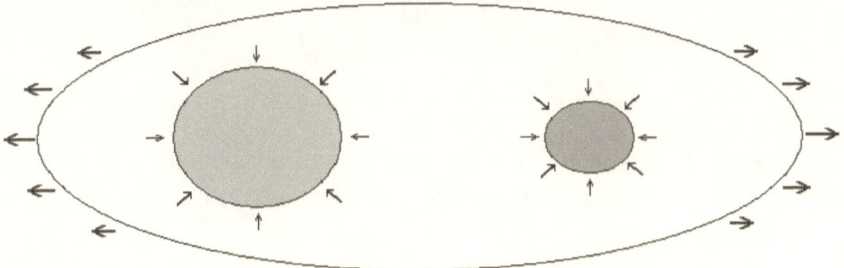

When dealing with space expansion, the flow is outward, away from the center of the object's mass. The expansion acceleration from the moon applies to both sides of the Earth across its diameter in line with the centers of the moon and the Earth. Thus, the accelerations from the expansion of space caused by objects in close proximity will enhance each other in a direct line between the centers of the two objects. This is what we saw in the balloon experiment in the chapter The Cause of Tides. As the balloon expanded the tension increased more in the line connecting the two objects than it did on the perpendicular axis. The increased tension of the balloon represents the distortion in space caused by the expansion of space.

This is a new law of physics never before understood. I would like to think of it as The Law of Tides. Remember that there is no great force or pressure on the Earth to cause the tides. The shape of objects in space is determined by the structure of space. The shape of the Earth conforms to the shape of space around it. The structure of space is determined by the contraction of space and the expansion of space combined. Therefore the tides are simply a result of the structure of space.

If we had actually used Equation 34 – Amplitude via Tidal Acceleration to do the calculation to obtain the amplitude of the high tide, we would have obtained the same result as from Butikov's equation. We will see why by making some adjustments to Equation 34. Let's replace the tidal accelerations in Equation 34 with their equivalent expressions on the right side of the equation. This little adjustment will give us the following equation:

$$a_t = \frac{3}{2} * \frac{\dfrac{GM\ (moon)}{pc_x * (distance\ from\ Earth)^3}}{\dfrac{GM\ (Earth)}{pc_x * (radius\ of\ Earth)^3}} * radius\ of\ Earth$$

To simplify this equation we can factor out all the constants, like the gravitational constant and the length of a short parsec. Doing so results in the following equation:

$$a_t = \frac{3}{2} * \frac{\dfrac{mass\ of\ moon}{(distance\ from\ Earth)^3}}{\dfrac{mass\ of\ Earth}{(radius\ of\ Earth)^3}} * radius\ of\ Earth$$

The above equation is identical to Equation 31, which was discussed at the beginning of this chapter. Here we see once again that the structure of space is related to the masses of the objects under consideration divided by their distance cubed. But now we understand why the tidal effect diminishes with distance cubed. It's because the tidal effect is the combination of two different accelerations. Gravitational acceleration diminishes with distance squared and acceleration from space expansion diminishes with distance. The natural consequence is that the tidal effect diminishes with distance cubed. Once again there is no force that causes the tides. There is no force in this equation only masses and distances.

Let's make one more adjustment to the equation above for simplification which yields the following equation:

Equation 35 – Amplitude of Earth's Tides

$$a_t = \frac{3}{2} * \frac{mass\ of\ moon}{mass\ of\ Earth} * \frac{(radius\ of\ Earth)^4}{(distance\ from\ Earth)^3}$$

Equation 35 is the same as Butikov's which is shown at the beginning of this chapter in Equation 30. We're back to where we started; but this same equation was derived in a different way. Because the equations are the same what differences does it make? The answer to that question comes down to understanding how the equation was derived. Space expansion provides a reasonable explanation of why the tides are high on both sides of the Earth. The arguments given to justify these tides with the gravitational differential across the Earth fail to explain the tides on both sides of the Earth; these current explanations do not hold up to critical scrutiny. Now let's understand what this equation means. This is an equation that allows us to compute the amount of restructuring an object in space will make when in close proximity to another object.

Formal Law and Equation

It's too early to declare this hypothesis a law. In order to be a law the measurements of the tides would have to be accurate enough to validate the equation. However, since the title of this chapter is The Law of Tides, I feel obligated to make some formal statements and rewrite the equation in a more formal way. The equations we have been using have applied to the tides on the Earth. The equation needs one final adjustment to be applicable to any two bodies in close proximity in space.

The Law of Tides:
1. Tides are the result of the restructuring of space as massive objects come into close proximity.
2. The restructuring of space around massive objects occurs as a result of the expansion and the contraction of space.
3. This restructuring of space also affects the shape of the objects in close proximity.
4. When two objects come into close proximity, both objects are stretched into ellipsoids, which produces tides.

Equation 36 – Amplitude of Tides, can be used to calculate the amplitude of the high tide on any object.

Equation 36 – Amplitude of Tides

$$a_t = \frac{3mr^4}{2Md^3}$$

Where:

a_t	is the amplitude of the high tide on object (M).
M	is the mass of the object under consideration.
m	is the mass of an object in close proximity.
r	is the radius of object (M).
d	is the distance between the two objects, M and m.

Diminution with Distance

We have witnessed an intimate relationship between the contraction and the expansion of space. The contraction of space always flows inward toward matter, whereas the expansion of space flows outward from matter, which is why we see its effects most clearly surrounding matter. We have seen two effects of expanding space that occur between two objects, the acceleration of the objects and the tides on objects in close proximity.

The acceleration of objects due to space expansion diminishes over distance, whereas the acceleration of objects due to space contraction diminishes over distance squared. So the effect of space contraction holds sway at close range, while the effect of space expansion dominates in the vast regions of space.

As we have seen in this chapter, the tides are caused by the distortion of space surrounding two objects in close proximity. This distortion is the result of space contraction and space expansion; because of this the tidal effect diminishes very rapidly with distance cubed. This makes sense because the effect of gravity diminishes over distance squared and the effect of space expansion diminishes over distance.

Tides Explained

In this chapter we have seen and made adjustments to equations to calculate the amplitude of high tides. One of the equations (Equation 32) utilized the ratio of accelerations caused by space contractions. Then we presented another equation (Equation 34) that included accelerations from space expansion and space contraction combined. Not only does this equation produce identical results, it explains more clearly why we have tides and why the tides diminish with distance cubed.

Space contraction holds a planet in its most compact form – a spheroid. Space expansion reconfigures the surrounding space of objects in close proximity into an ellipsoidal shape. The shape of objects in space conforms to the resulting structure of space. This is why we have high tides on both sides of the Earth.

Chapter 18

The Roche Limit

We all need to know our limitations.

Roche's Limit

French astronomer Edouard Roche (1820-1883), calculated the limit at which the tidal forces of a moon's parent planet would exceed the moon's internal binding force and tear it into smaller pieces. That is believed to be the reason why some planets have rings. Large objects coming too close to a planet will be shredded into small pieces which eventually form a ring around the planet. The rings around Saturn and other planets are below the Roche limit. The equation for the Roche limit is given below and pertains to satellites held together by their own gravitational attraction:[29]

Equation 37 – Roche Limit

$$R_L = 2.456 * \text{radius of planet} \left[\frac{\text{density of planet}}{\text{density of satellite}} \right]^{1/3}$$

Io is very close to Jupiter at a distance of 421,600 kilometers. How close would Io need to get to Jupiter to become a ring? The density of Jupiter is 1240 kg/m^3, and its radius is 71.49x10^6 meters. The density of Io is 3530 kg/m^3.

Let's calculate the Roche limit for Jupiter and Io:

$$R_L = 2.456 * 71.49\text{x}10^6 \left[\frac{1240}{3530} \right]^{1/3}$$

$$R_L = 123,886 \text{ kilometers}$$

The Roche limit for Io is about 123,886 kilometers. So, right now, Io is safe at its current distance of 421,600 kilometers.

Now let's use Equation 36 – Amplitude of Tides, to calculate the amplitude of the tidal bulge on Io at the Roche limit. We'll use the Roche limit as the distance between objects in close proximity.

$$a = \frac{3}{2} * \frac{1.899 \times 10^{27}}{8.932 \times 10^{22}} * \frac{(1.821 \times 10^{6})^4}{(1.238 \times 10^{8})^3}$$

$$a = 184 \text{ kilometers}$$

If Io were to approach the Roche limit the amplitude of its tidal bulge would be about 184 kilometers. Is this the point at which the satellite is crushed into pieces? My suspicion of the Roche limit was aroused when I first noticed the Roche limit equation contained the cube root of a ratio of densities. This is somewhat similar to the equation used to calculate the amplitude of tides. This gave me the uneasy feeling that the Roche Limit was derived under the assumption that the tidal force was the reason why a large satellite could be rendered into billions of tiny bits to make a beautiful ring around a planet.

It might seem logical, after all, if a moon were to keep stretching into a more eccentric elliptical shape, eventually it would break - right? I don't think so, but that is what the Roche limit suggests. There are a couple of problems with this line of reasoning and the first is this: The tidal force is assumed to be the culprit in the derivation of the equation that is used to convict the culprit. The circumstantial evidence fits, i.e. the rings around the planets are all within the Roche limit, therefore the verdict arrived at is – the tidal force is guilty as charged.

Here is the second problem: the assumption that the tidal force is the cause of the demolition of satellites is in direct conflict with the previous analysis of structured space. In the explanation of the cause of tides we made statements such as: "The mass of an object conforms to the space provided and fills the space to its lowest energy level." We also stated that no massive forces need be applied. If no forces are applied in the stretching processes there can be no rendering apart. The mass simply fills the space provided.

The Limit of Stability

We need to put the Roche limit to the test. Let's calculate the gravitational acceleration on the surface of Io due to its own mass. Then we'll calculate the gravitational acceleration on Io caused by Jupiter at the Roche limit.

$$a_i = \frac{G * \text{mass of Io}}{(\text{radius of Io})^2} \qquad a_j = \frac{G * \text{mass of Jupiter}}{(\text{Roche Limit})^2}$$

$$a_i = \frac{G * 8.932 \times 10^{22}}{(1.821 \times 10^6)^2} \qquad a_j = \frac{G * 1.899 \times 10^{27}}{(1.238 \times 10^8)^2}$$

$$a_i = 1.79 \text{ m/s}^2 \qquad a_j = 8.27 \text{ m/s}^2$$

What we can conclude from this is that at the Roche limit the gravitational acceleration of Jupiter on Io is 4½ times stronger than Io's own gravitational field. At this point, Io could not possibly hold itself together. So, it's not the tidal "force" that destroys a satellite, it's the "force" of gravity. When a planet consumes more space at the surface of a satellite than the satellite itself, then the gravitational stability of the satellite is compromised.

We can logically say that the stability of a satellite held together by its own gravitational attraction is compromised when another gravitational attraction from a nearby object is equal to or greater than its own. Let me express this again in a different way. A satellite will become unstable when the gravitational attraction from a nearby object can start attracting mass from the satellite to itself. At that point, matter will begin to float off the surface of the satellite toward the object of greater attraction. Note that this applies to all objects having a close encounter with a more massive object. The object need not be a satellite.

We can express this mathematically by using the gravitational acceleration formula. Let a_s represent the acceleration at the surface of the satellite due to its own gravitational attraction. Let a_p represent the acceleration at the surface of the satellite due to the gravitational attraction of the planet it orbits. The two equations are shown below:

$$a_s = \frac{G * \text{mass of satellite}}{(\text{radius of satellite})^2} \qquad a_p = \frac{G * \text{mass of planet}}{(\text{distance from planet})^2}$$

When these two accelerations are equal, the satellite becomes unstable. Setting the expressions for these two accelerations equal to each other we get the following equation:

$$\frac{G * \text{mass of satellite}}{(\text{radius of satellite})^2} = \frac{G * \text{mass of planet}}{(\text{distance from planet})^2}$$

Solving for the distance between the planet and its satellite we get an equation that gives us the limit of stability:

Equation 38 – Limit of Stability

$$L_s = R_s \sqrt{\frac{M_p}{M_s}}$$

Where:

L_s Limit of Stability – distance between a satellite and planet
R_s Radius of satellite
M_p Mass of planet
M_s Mass of satellite

This equation gives us the distance between a planet and another object at which the gravitational accelerations will be equal. At that distance the smaller object will become gravitationally unstable. Let's apply this equation to Io and Jupiter to find the limit of stability.

The mass of Jupiter is 1.899×10^{27} kilograms. The mass of Io is 8.932×10^{22} kilograms, and the radius of Io is 1.821×10^{6} meters.

$$L_s = 1.821 \times 10^{6} \sqrt{\frac{1.899 \times 10^{27}}{8.932 \times 10^{22}}}$$

$$L_s = 265,520 \text{ kilometers}$$

This is the distance from Jupiter at which Io would become unstable and begin to fall apart. This is more than twice the distance of the Roche limit – 123,886 kilometers. At this distance (265,520 kilometers) the amplitude of Io's tides would only be about 18.7 kilometers. The tidal effect and the structure of space are vindicated. The real culprit is gravity. It's gravity, the contraction of space, that causes the satellite to break up.

Yes, the satellite would take on an ellipsoidal shape due to the tidal effect, but this would not result in the destruction of the satellite. The restructuring of a large mass may weaken the electromagnetic bonding, but this is not the cause of instability. The limit of a satellite's stability is when the planet's gravitational tug is greater than that of the satellite at the satellite's surface. The distance at which a satellite becomes unstable is greater than the Roche limit.

What we've discussed here is the gravitational effect of large objects held together by their own gravitational attraction that come into close proximity. Artificial satellites and spacecraft are held together by electromagnetic forces at the molecular level so are unaffected by the "stress" of gravity or tidal effects. Also, the gravitational "force" from planets across small objects would be too small to have any substantial affect. Of course, falling into a black hole would be a problem.

The Fundamental Force

Figure 36 – Jupiter and Io, illustrates the relative sizes of Io and Jupiter. Look closely, the black dot on Jupiter is the shadow of Io.

Figure 36 – Jupiter and Io

Courtesy of NASA/JPL/University of Arizona

http://photojournal.jpl.nasa.gov/catalog/PIA02860

Chapter 19

Planets in Precession

Around and around we go.

Precession Explained

One very interesting phenomena that occurs in solar systems is that of perihelion precession. Planets revolve around their sun in slightly elliptical orbits. The perihelion is the point of orbit closest to the Sun. The aphelion is the point of orbit farthest from the Sun. One word, "apsides" applies to both. These points advance along in the orbits of all planets with elliptical orbits. A perihelion precession is the slow rotation of the elliptical orbit of a planet around its sun, which is located at the focus nearest the perihelion of that planet's orbit; thus the term perihelion precession. Because the orbit precesses, the term orbital precession is also often used. The point of rotation is the focus of the orbit where the Sun is located. After each orbital cycle the planet's position will have moved slightly forward from its position on the previous cycle.

Figure 37, although highly exaggerated, illustrates the precession of apsides. Figure 37 shows the orbital path of a planet revolving around the Sun. The planet is pictured at the aphelion of each cycle. After each cycle the planet's position rotates counterclockwise. In our example it's easier to illustrate the position of a planet at the aphelion than at the perihelion, so in our example we are visualizing the aphelion precession. The perihelion precesses as well but is much more difficult to illustrate and visualize.

One of the most interesting and talked about precessions is that of the planet Mercury. Mercury's orbital precession came into prominence when Einstein announced that the previously unaccounted for precession of Mercury's orbit was the result of the spacetime distortion which he explained through his theory of relativity.

What is the cause of precession? This chapter begins the investigation as to the real causes of precession. We will soon see what can and what cannot cause a perihelion precession – and why.

Figure 37 – Perihelion Precession

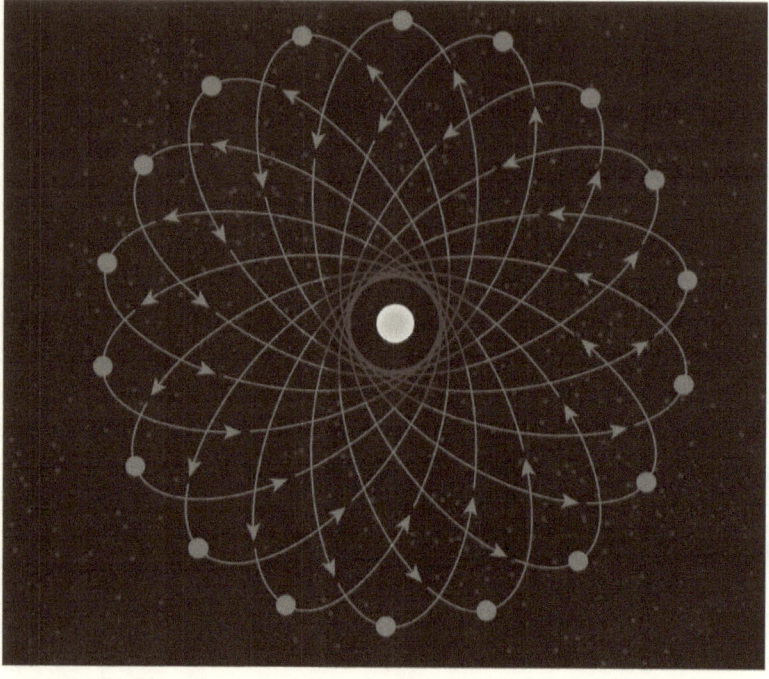

Courtesy of Brandon Fall

Mercury's Perihelion Precession

Before answering that question we need to understand the units of measurements for precession. Precession is measured in arc-seconds. An arc-second is the measure of a tiny portion of the circumference of a circle. In this case the circle is the orbit of planets. Envision the circumference of a circle divided into 360 equal parts. One section would be one degree of arc. Divide that into 60 parts to get 60 arc-minutes. Divide an arc-minute into 60 more parts to get 60 arc-seconds.

Now back to the question: what causes perihelion precession? The blame for perihelion precession is placed on the perturbations caused by neighboring planets. Planets have a gravitational effect on each other. The effect is to draw both planets slightly off course and closer to each other. Perturbation is the word used when a planet is drawn off course by another planet because it's normal orbital path is perturbed. We will examine the mechanics of perturbations more deeply in the next chapter.

532 arc-seconds per century of the advancement of Mercury's perihelion precession have apparently been accounted for via perturbations. That was fine until 1859, when French mathematician and astronomer Urban LeVerrier discovered that the precession of Mercury's perihelion was actually rotating at 575 arc-seconds per century - 43 arc-seconds per century faster than expected.

LeVerrier is the same scientist who in 1846 discovered the planet Uranus to be off course. Uranus was perturbed by a yet undiscovered planet. Shortly after LeVerrier calculated its expected location, the planet Neptune was found. After LeVerrier discovered the unaccounted for 43 arc-second advance in Mercury's perihelion, many possible explanations were presented to account for this behavior. One theory was that another planet was orbiting the Sun inside the orbit of Mercury. The expectation of finding another planet was so great that it was given a name – Vulcan. The planet was never found, and the 43 arc-second advance remained a mystery until 1915 when Einstein developed the General Theory of Relativity.

Einstein was able to use his field equations to show an advance of 43 arc-seconds per century in Mercury's perihelion. The precession is caused by a space-time distortion in the vicinity of the Sun. Remember that mass causes a distortion of space-time. Time runs slower when closer to a massive body than when farther removed. At its perihelion, Mercury moves faster than expected because time is flowing slower. This additional acceleration causes the perihelion to advance 43 arc-seconds per century; giving much credence to the theory of relativity. This phenomenon of precession of apsides is now well known because of Einstein's General Theory of Relativity.

The precession described above is from the effect of the time-space distortion caused by massive objects. Einstein believed that the distortion of time and space is the cause of gravity, and this also has an effect causing the precession of the planets' orbits. This effect applies to all planets, but diminishes over distance squared, just like gravity. Currently the observed precession of the three inner planets due to the effect of relativity is shown in Table 21:[30]

Table 21 – Precession of Inner Planets

Mercury	43.1 ± 0.5
Venus	8.4 ± 4.8
Earth	5.0 ± 0.8

Table 22 shows Mercury's perihelion precession in arc-seconds per century, as well as the cause of the precession. Notice that the two precessions add up to the observed precession. Everything seems to fit quite nicely.

Table 22 – Precession of Mercury's Perihelion

Precession from Perturbations by Planets	532
Precession from effect of Relativity	43
Observed Precession of Mercury	575

Calculations and Observations

We can validate the precession of a planet from astronomical observations by using Kepler's Law of Periods. In order to make this validation we would need a very accurate measurement of the planet's period of orbit and distance from the Sun. We would do well to use the planet Mercury for our test case because it's orbit is more eccentric than the others, and it is closest to the Sun. The eccentric orbit makes it easier to measure the planet's period of orbit, and its close proximity to the Sun will give us the most accurate measurement of the radius of orbit. Mercury also has the greatest amount of precession so it is the best possible test case.

Pluto, although very eccentric is too far from the Sun to get an accurate measurement of its period of orbit or its radius from the Sun.

Remember Table 3 – Kepler's Law of Periods in the chapter on Kepler's Laws which compared the period and radius of the planets' orbits. The discrepancy between the two grows larger as the distance from the Sun increases.

The period of orbit used in these calculations is the sidereal period. It is the amount of time it takes a planet to make a complete revolution around the Sun, and return to exactly the same point in relation to the stars. This is the most accurate measurement possible of a planet's orbital period.

We saw from the explanation of Kepler's Law of Periods that if we measure a planet's orbital radius in AUs and its orbital sidereal period in years, then we have an expression to equate the two. That expression is: $T^2 = D^3$. That is, the time of a planet's orbital period squared is equal to the orbital radius cubed. Let's take the sidereal time of Mercury's period, obtained from the NASA website[31]. The sidereal orbital period of the Earth is 365.256 days. The sidereal orbital period of Mercury is approximately 87.969 days.

Before making any calculations allow me to make a minor adjustment to Mercury's period. I'd like to increase it by just 3.427974778 seconds. By making this minor adjustment, Mercury's period goes from approximately 87.969 days to exactly 87.969039675634 days. Remember we are about to calculate, with the highest degree of accuracy possible, the precession of Mercury's perihelion over a 100 year period. We made this change to be more precise; since we need to be more accurate than just approximation. We'll also need the accuracy of as many decimal digits as possible.

Now let's divide Mercury's orbital period by the Earth's orbital period to get the sidereal period of Mercury in years.

We get: 87.969039675634 ÷ 365.256 = 0.240842148180.

This is the astronomically observed sidereal period of Mercury in Earth years.

The given radius of Mercury (Semimajor axis in AUs) is also taken from the same NASA webpage. This is already in the form we needed - 0.38709893 AUs. We now have two measurements, Mercury's orbital radius and orbital period. These should match up exactly according to Kepler's third law. And now with Kepler's assistance let's cube the radius and then take the square root. This will give us the calculated orbital period in years that we would expect to see from astronomical observation.

Orbital Period = $\sqrt{(0.38709893)^3}$ = 0.240842405538

The difference between the calculated and the observed orbital period is:

0.240842405538 - 0.240842148180 = 2.5736×10^{-7}

There is a discrepancy albeit a very slight one. The results calculated in accordance with Kepler's laws do not match the astronomical observations. Multiply the difference by the number of seconds in a year (3.15581184×10^7) and the product will be the number of seconds that Mercury completed its orbit before its expected completion time:

$2.5736 \times 10^{-7} * 3.15581184 \times 10^7 = 8.1217$ seconds

This is 8 seconds faster than Kepler's Law will allow. Either Kepler's Law is only a very close approximation, or there is some other reason for Mercury's early arrival. A scientific law is called a law for a good reason. It's something that works all the time, every time, exactly. There is a reason for the early arrival of Mercury, and that reason is called precession. Mercury is getting an assist from another source to give it a little more acceleration. With an increase of acceleration there is also an increase of velocity, thus giving the planet a slightly faster cycle time.

Calculating Precession

Now we can calculate the precession of Mercury's orbit in arc-seconds per century. First we need to get the difference between the two velocities. We'll calculate the accelerated velocity and then subtract the expected velocity from it. From that we can calculate the precession. First we need the orbital periods converted to seconds. This is easily accomplished by the multiplying the orbital period given in years by the number of seconds in a year - 31558118.4. The orbital periods in years were already calculated in the previous section. The calculations to obtain the periods in seconds follow:

Mercury's calculated orbital period in years is 0.2408424055.
Mercury's calculated orbital period in seconds is:
    0.2408424055 * 31558118.4 = 7,600,533.1497

Mercury's observed orbital period in years is 0.2408421482.
Mercury's observed orbital period in seconds is:
    0.2408421482 * 31558118.4 = 7,600,525.0280

The difference between these two time periods:
    7,600,533.1497 − 7,600,525.0280 = 8.1217 seconds.

This is the number of seconds Mercury arrived early.
We can calculate the velocity of Mercury by the dividing the circumference of orbit by the orbital period as in the equation below. We'll calculate twice using the two orbital periods just obtained.

$$\text{Velocity} = \frac{2\pi \text{ orbital radius}}{\text{orbital period}}$$

This is the normal velocity based on Mercury's observed radius:

$$\text{Velocity} = \frac{2\pi \ 57,909,175,675}{7,600,533.1497} = 47,872.178778$$

This is the fast track velocity from Mercury's observed orbital period:

$$\text{Velocity} = \frac{2\pi \ 57,909,175,675}{7,600,525.0280} = 47,872.229933$$

The difference in velocities is:
    47,872.229933 − 47,872.178778 = .051155 meters/second

The increased velocity given to the Planet Mercury is only 51 millimeters per second. That's about 2 inches very second. The velocity difference we have just calculated is the difference between the observed distance of Mercury from the Sun and the observed sidereal period of Mercury's orbit. We calculated the velocity using two different observations. We should have been able to calculate Mercury's exact velocity both ways and acquired the same results. We did not. If the observations are correct and Kepler's third law is in fact a law, then there must be an increase in the planet's velocity from another source in order to speed up Mercury's arrival time.

We can now calculate the precession rate based on this velocity difference. Multiply this velocity by the number of seconds in a century. This gives us the extra distance Mercury traveled in 100 years.

$$.051155 * 31558118.4 * 100 = 1.61436 \times 10^8 \text{ meters}$$

Divide that by the circumference of Mercury's orbit to get the ratio of extra distance traveled compared to the orbit.

$$1.61436 \times 10^8 \div (2\pi * 57,909,175,675) = 4.43683 \times 10^{-4}$$

Multiply that ratio by 360 to get the number of degrees and then multiply by 60 times 60 to get then number of arc-seconds.

$$4.43683 \times 10^{-4} * 360 * 3600 = 575 \text{ arc-seconds per century}$$

575 is the number of arc-seconds of precession per century that have been observed in Mercury's perihelion. This was the justification for the minor adjustment to Mercury's sidereal period. That little adjustment allows us to see exactly what astronomers have seen while observing Mercury's perihelion advance.

We have just validated the 575 arc-seconds of precession per century that astronomers have observed. These calculations were made from measurements of Mercury's sidereal period and mean distance from the Sun. We calculated a discrepancy of 8.1217 seconds in Mercury's period, which indicates an additional assist in Mercury's acceleration. These calculations were made from astronomical observations and confirm the

575 arc-seconds of precession. But, these observations do not identify the cause of the additional acceleration which causes the precession. The major cause of precession has been attributed to planetary perturbations, and a minor cause has been attributed to relativity.

Precession of Planets

We still need to consider the effects of expanding space in our own solar system. The expansion rate is very small at the local level but the effects can still be measured. We can see the tides. We can also see the effect of additional acceleration on the Pioneer spacecraft. One effect of expanding space that may be difficult to observe is the precession of apsides on all the planets.

The precession caused by expanding space has not yet been realized by the scientific community. We have already seen the rotation of galaxies caused by the expansion of space. The additional velocity of rotation in galaxies can be considered a precession in rotation. The same thing happens within the solar system only on a much smaller scale. There is a small increase in acceleration on all planets in the solar system due to the expansion of space. This increase in acceleration is a cause of the precession of their orbits. The precession of a highly circular orbit is not easily detected, whereas the orbital precession of a planet with an elliptical orbit – like Mercury can be more easily observed.

This same phenomena, on an even smaller sale, would also occur in the moons circling the planets. So on the large scale we have the precession of galaxies because of the large increase of space expansion. Within the galaxies are solar systems with planets whose orbits are in precession around them. The planets have moons whose orbits are also in precession. The smaller the distance, the smaller the precession.

Table 23 below shows the acceleration for each planet due to gravity and the additional acceleration due to space expansion. The "acceleration" column is the normal acceleration which is calculated from the Newtonian acceleration equation. The "X Accel" column shows the planet's acceleration due to the expansion of space. This was calculated from the expansion equation given in the chapter on Space Expansion. Notice the difference in magnitude of expansion acceleration compared to gravitational acceleration. The acceleration caused by expanding space is significantly smaller than its gravitational counterpart.

The "V1" column shows the planet's velocity according to Newtonian dynamics. The "V2" column shows the planet's velocity resulting from both accelerations. The data in the "V2" column was obtained by adding the two accelerations, multiplying the result by the radius of the planet's orbit, and then taking the square root of that result. The calculation for Mercury's velocity is shown below.

$$V2 = \sqrt{ad} = \sqrt{(3.9586 \times 10^{-2} + 7.8262 \times 10^{-8}) * 5.7909 \times 10^{10}}$$

$$V2 = 47{,}878.85 \text{ meters / second}$$

Table 23 – Acceleration of Planets

Planet	Radius	Acceleration	X Accel	V1	V2
Mercury	5.7909E+10	3.9586E-02	7.8262E-08	47,878.81	47,878.85
Venus	1.0821E+11	1.1337E-02	4.1883E-08	35,025.59	35,025.65
Earth	1.4960E+11	5.9318E-03	3.0295E-08	29,788.89	29,788.96
Mars	2.2794E+11	2.5551E-03	1.9883E-08	24,132.92	24,133.02
Jupiter	7.7841E+11	2.1909E-04	5.8222E-09	13,059.07	13,059.24
Saturn	1.4267E+12	6.5216E-05	3.1765E-09	9,645.99	9,646.22
Uranus	2.8710E+12	1.6106E-05	1.5786E-09	6,799.90	6,800.23
Neptune	4.4983E+12	6.5606E-06	1.0075E-09	5,432.44	5,432.86
Pluto	5.9064E+12	3.8053E-06	7.6732E-10	4,740.85	4,741.33

There is a very small increase in the each of the planet's velocities because of the expansion of space surrounding the Sun. The velocity increase is shown in the "Vx" column of Table 24 – Precession of Planets. Table 24 is an extension of Table 23 – Acceleration of Planets. These velocities are in meters per second and can be attributed to acceleration from space expansion. Multiply the velocity of Mercury, in the "Vx" column, by 1000 and we can see that the additional velocity of Mercury is about 47 millimeters per second. That's less than 2 inches per second. That's not very much at all, but it can be measured by the shortened amount of time it takes Mercury to revolve around the Sun.

Table 24 – Precession of Planets

Planet	Radius	Vx	d / century	% of orbit	Precession
Mercury	5.7909E+10	0.04733	1.4936E+08	4.1049E-04	532.00
Venus	1.0821E+11	0.06470	2.0417E+08	3.0029E-04	389.18
Earth	1.4960E+11	0.07607	2.4006E+08	2.5540E-04	331.00
Mars	2.2794E+11	0.09390	2.9632E+08	2.0691E-04	268.15
Jupiter	7.7841E+11	0.17352	5.4760E+08	1.1196E-04	145.10
Saturn	1.4267E+12	0.23492	7.4135E+08	8.2700E-05	107.18
Uranus	2.8710E+12	0.33324	1.0516E+09	5.8298E-05	75.55
Neptune	4.4983E+12	0.41711	1.3163E+09	4.6574E-05	60.36
Pluto	5.9064E+12	0.47796	1.5083E+09	4.0644E-05	52.67

We can multiply that additional velocity in the "Vx" column by the number of seconds in a century ($3.15576e^9$) and get the number of additional meters traveled in that century. That calculation is done below. The "d / century" column shows exactly that.

$$0.04733 * 3.15576 \times 10^9 = 1.4936 \times 10^8 \text{ meters per century}$$

Divide that distance by the circumference of the planet's orbit to find the percentage of the orbit attributed to this velocity. The result of that calculation is shown in the first row of the "% of orbit" column. That calculation for Mercury is shown below:

$$1.4936 \times 10^8 \div 2\pi \, 5.7909 \times 10^{10} = 4.1049 \times 10^{-4}$$

We need to take just a few more steps to calculate the precession of Mercury's perihelion in number of arc-seconds per century. Multiply the "% of orbit" by 360 to get the number of degrees; and multiply that by 60 to get the number of arc-minutes, and then multiply by 60 again to get the number of arc-seconds. This is the precession of the perihelion that can be attributed to space expansion. The "Precession" column shows the precession for each planet in arc-seconds per century. The calculation for Mercury is shown below:

$$4.1049 \times 10^{-4} * 360 * 60 * 60 = 532 \text{ arc-seconds per century}$$

We can see from this table that the precession of the planets' orbits decreases with distance from the Sun.

One other point of interest. It's because of this slight precession of each planet's orbit that throws the calculations for Kepler's Law of Periods slightly off their expected value. The sidereal period squared and the orbital radius cubed will never match up exactly because of the precession from space expansion. It's not that Kepler's Laws are incorrect, it's just that another acceleration makes Kepler's Law appear to be slightly off. Kepler's Law of Periods is correct but applies only to space contraction.

Perturbation or Space Expansion

Table 22 – Precession of Mercury's Perihelion, shows the arcs-seconds in Mercury's perihelion precession. Table 25 – Precession of Mercury, is basically the same table with an additional row that shows what has been calculated for the perihelion shift due to the expansion of space. All the advances of precession should add up to what is observed. They do not. Something is obviously terribly wrong. There's one too many players in this game. There are now two claims for the 532 arc-seconds of precession. They both cannot be right. One is fraudulent and must be eliminated. This is a dilemma! Which is the culprit – the one really causing the precession? And which is the imposter only pretending to cause the precession?

Table 25 – Precession of Mercury

Precession from Perturbations by Planets	532
Precession from Space Expansion	532
Precession from effect of Relativity	43
Observed Precession of Mercury	$\neq$ 575

By proposing an additional perihelion shift due to expanding space we have muddied the clear waters of the science of apsidal precession. Everything added up just right before, but now it doesn't. This might be sufficient evidence for some to suggest that space is not expanding, or if it is, that it has no effect on apsidal precession. Before making any rash judgments let's consider the workings of perturbations. Because if the perturbations from other planets can cause a precession or a recession of Mercury's apsides, we still cannot explain why the precessions and recessions don't total up to what is observed. In order to resolve this dilemma we will focus on perturbations. Then we'll come back to Mercury's perihelion precession.

Chapter 20

Planetary Perturbations

Even planets can be perturbed.

Perturbed Planets

Astronomical tables of the planets' orbits were published as early as 1821. Subsequent observations of Uranus found it to be slightly off its expected course. In 1846 French mathematician and astronomer Urbain LeVerrier deduced that another planet must be pulling Uranus off course. When one planet pulls another off course it's called a perturbation. With Uranus off course, the search for a new planet resulted in the discovery of Neptune just where Le Verrier had predicted, making Neptune the first planet discovered based on mathematical calculations.

Let's calculate the gravitational attraction Neptune has on Uranus when they come into conjunction. The mass of Neptune is 1.0244×10^{26} kilograms. The distance between the two planets is 1.62728×10^{12} meters during their conjunction. The calculation is shown below:

$$a = \frac{6.6742 \times 10^{-11} * 1.0244 \times 10^{26}}{(1.62728 \times 10^{12})^2} = 2.5819 \times 10^{-9} \text{ meters/sec}^2$$

The acceleration of Uranus toward Neptune is 2.5819×10^{-9} meters / sec^2. This is very small compared to the Sun's gravitational attraction on Uranus of 1.61×10^{-5} meters / sec^2. We can take the ratio of these two accelerations and multiply by the distance Uranus is from the Sun (2.871×10^{12} meters). This will give us a good approximation of how far Uranus is lured off course by Neptune.

$$(2.5819 \times 10^{-9} \div 1.61 \times 10^{-5}) * 2.871 \times 10^{12} = 460,412 \text{ kilometers.}$$

From this calculation we see that Neptune pulls Uranus off course by about 460,000 kilometers, or about 286 thousand miles. This may seem like a lot, but not when compared with the eccentric size of Uranus's focus which is 294 times greater. Uranus's diameter is about 51,000 kilometers, so at worst Uranus only strays off course about 9 times its own diameter. When observing Uranus through a telescope that's not really very much, but a keen observer like Urbain LeVerrier noticed it.

The same calculations were performed to find the affect Jupiter has on the Earth. When these two planets come into conjunction, Earth strays off course by 8,082 kilometers – about 5,023 miles. That's more than the Earth's radius, and though it may be interesting, I never even noticed. There may be many other things that can perturb us more quickly.

Planets can perturb the orbits of other planets, but in most cases it's like driving on the highway in a car and going over a little hill, or dipping into a little valley. The overall time and speed are only slightly affected and the car remains on course. It's the same with planets and they remain on course in elliptical orbits. Planetary orbits can be predicted many years in advance. Perturbations by other planets may cause a slight and temporary wandering, but the planet pulled off course will return to its calculated path.

When we speak of one planet perturbing another we are really talking about two planets perturbing each other. The planet closer to the Sun is in the inferior orbit, while the planet farther out is in the superior orbit. Their respective paths are distorted by a mutual attraction. When coming into conjunction, the planet in the superior orbit will shorten its orbit and increase in velocity. At the same time, the planet in the inferior orbit will increase its orbit and decrease its velocity. Between the two planets their combined angular momentum will be preserved. After passing, both planets will return to their normal orbits. During this exchange all laws of physics are obeyed.

Perturbation of Orbits

For almost 150 years astronomers have attributed 532 arc-seconds of precession per century in Mercury's orbit to perturbations by other planets. Might they be wrong? Let's carefully consider the following arguments. In the last chapter we saw that in order to experience a precession of apsides a planet must have an increase of acceleration for a bit more velocity. A little assist in acceleration from another source other than the Sun could possibly result in a precession of apsides. The word "possibly" is used in the previous sentence because at this point we cannot say with certainty that all assists in acceleration will cause a precession of apsides.

Let's use Mercury and Venus as an example and discuss their interaction in greater detail. The objective is to see if perturbations of Mercury's orbit can cause a precession. As Mercury and Venus come into conjunction, Mercury's gravitational effect will draw Venus in closer to it and the Sun. This will give Venus an additional acceleration which will increase its speed. Also, because Venus is drawn in closer to the Sun, its orbit around the Sun is shortened. Combining the shortened orbit and the increase in velocity, Venus will complete its revolution around the Sun slightly ahead of its expected arrival time. Could this be the cause of precession of apsides? As a general statement we can say that planets in an inferior orbit will give planets in a superior orbit an assist in acceleration. This may possibly cause an apsidal *precession* in the superior planet's orbit, but that is yet to be determined.

Now let's take a look at the other side of the coin. Venus has a gravitational effect on Mercury. As they come into conjunction Mercury will be drawn very slightly toward Venus – away from the Sun. This will increase the length of Mercury's orbit, and decrease its velocity. Remember, as planets get farther away from the Sun their orbital speed diminishes. The increased orbital size and slower velocity will cause Mercury to arrive later than expected. This is the opposite of precession – let's call it a recession. Another general statement we can make is that all planets in superior orbits will cause a deceleration in planets of inferior orbit. This could possibly cause an apsidal *recession* in the inferior planet's orbit.

The perturbation calculations using Newtonian dynamics show that the gravitational attraction of Venus on Mercury draws it off orbit by about 188 km. Jupiter has a greater effect, moving Mercury 357 km off course. That's almost twice as much as Venus's effect on Mercury. But consider the radius of Mercury – 2,439 km. A perturbation of 357 km is just a fraction of the planet's radius. The combined effect of all the planets perturbing Mercury's path may cause a *recession* of apsides of Mercury's orbit, but not a *precession*.

Given this understanding and knowing that Mercury has the most inferior orbit of all planets, we can deduce that perturbations from the other planets cannot possibly cause a *precession* of Mercury's apsides. They can only possibly cause a *recession*! So there must be another reason for Mercury's acceleration assist and apsidal precession.

Let's take a look at some figures below to help us visualize what actually happens when one planet perturbs another. We'll be using Mercury and Venus again in our examples. But first let's look at the orbital paths of Mercury and Venus with no perturbations.

Figure 38 shows the orbital paths of both Mercury and Venus. Venus of course is the outer path which is very circular. Mercury appears to be very close to a circle. But we can now see how elliptical its orbit is because the Sun is positioned at one focus, which is considerably off center for Mercury and right on center for Venus.

Figure 38 – Orbits of Venus and Mercury

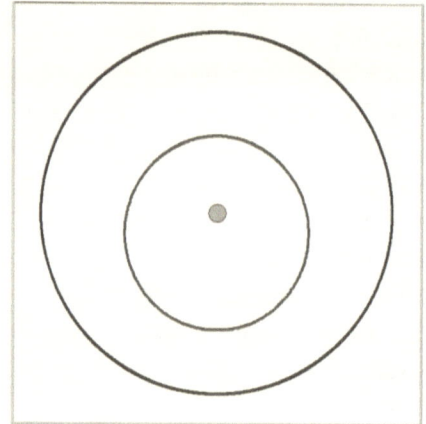

The following four figures show the perturbation of Mercury's orbit due to the gravitational attraction of Venus. In the various figures Venus is positioned at 90 degree angles around the Sun. Venus' position is stationary while Mercury makes a revolution around the Sun. The Sun is at the center of the diagrams. In each case we can see the whole orbit of

Mercury being displaced toward Venus. The displaced orbit is superimposed over the original unperturbed orbit. The orbital displacement is used to illustrate the location of Mercury no matter where it is located in its orbit. The orbital displacement has been exaggerated 30 thousand times in order to visually capture the direction of displacement. Otherwise the displacement would be too small to be seen.

Figure 39 – Venus to the Right

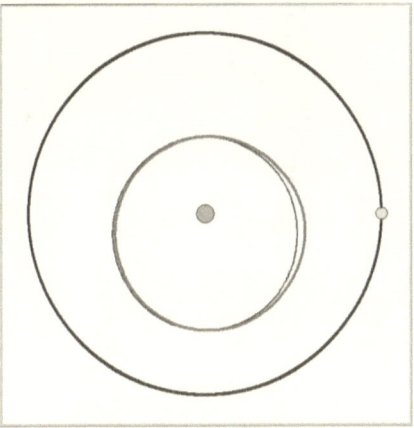

Figure 40 – Venus on Top

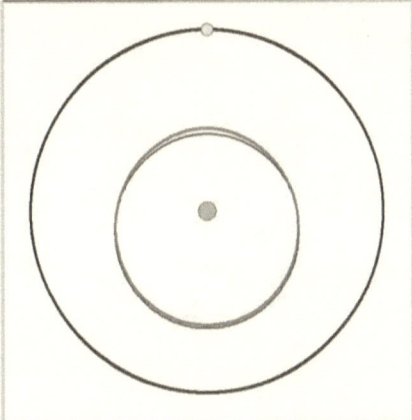

In Figure 40 the perturbation is not as great because Mercury is at its perihelion. This means Mercury is at its closest point to the Sun. In this position the Sun's pull is much stronger so the perturbation caused by Venus is less.

Figure 41 – Venus to the Left

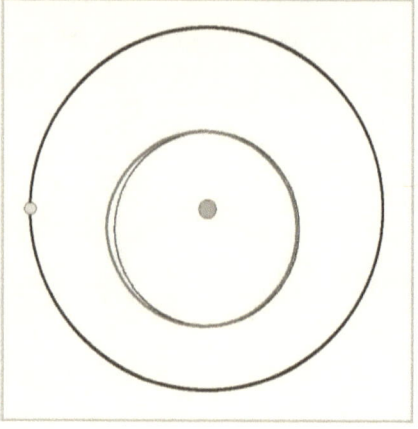

Figure 42 – Venus on the Bottom

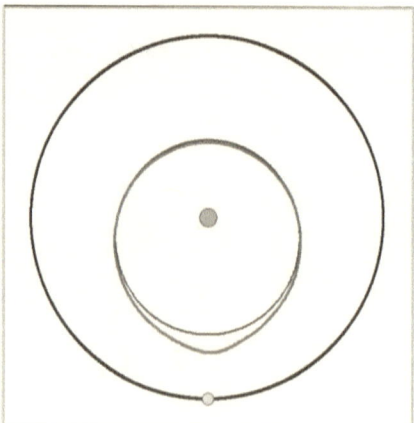

In Figure 42 Mercury is at its aphelion in relationship to the Sun. This is the point that Venus would have the greatest affect on the path of

Mercury and cause the largest displacement. Even so, we always see Mercury returning to its original path, with no advance of the perihelion.

Careful observation of these orbital displacements shows that the orbit is displaced but not turned. There is no indication that there would be a precession or recession of apsides. Envision the sequence of events, as pictured in Figure 39 – Figure 42, as Venus and Mercury revolve around the Sun. The whole orbit of Mercury is drawn closer to Venus, wherever Venus is located. We can see that the orbital path of Mercury gyrates around the Sun but never loses its orientation to the Sun. This means that nothing ever happens with perturbations to cause a perihelion precession. The effect of perturbations bring about what we might call a gyration in the orbit of Mercury, but not precession. Even after thousands of revolutions these perturbations cause no precession of apsides whatsoever.

The erroneous assumption is that if a planet gives another planet an assist in acceleration, it will cause an apsidal precession in the superior planet's orbit. This assumption is not true. Yes, a planet's path around the Sun may be shortened, and its velocity may be increased, and it may arrive ahead of schedule, but that will not cause an apsidal precession, or recession. Sure, the assisted planet may cycle in its orbit a little faster, but its orbit does not rotate. The path is perturbed, but not precessed, or recessed.

One hundred and fifty years ago astronomers observed and measured the precession of Mercury's orbit. They knew then that perturbations from other planets would slightly alter the period of a planet's orbit. They naturally assumed that a reduced cycle time would cause a precession in the planet's orbit. That reduced cycle time could only come from a planet in an inferior orbit. That's why there was a search for a planet closer to the Sun than Mercury. Astronomers were so sure of finding it they gave it a name – Vulcan; the planet was never found.

Once again take a look at Figure 39 through Figure 42. Recall that the offset of the perturbed orbits are highly exaggerated to enhance the visual effect. Notice that in each diagram some parts of the perturbed orbits are drawn closer to the Sun and some parts are drawn away from the Sun. Averaging out the perturbations over a 100 year period will yield a net difference in orbit of only a few hundred meters (that's meters, not kilometers), which means that perturbations cannot possibly cause precessions of any significance.

Newtonian dynamics do not show a precession resulting from perturbations. There is no law of perturbations that will show us a precession. Algorithms based on a false assumption were developed to show a precession. These assumptions have stood firm for 150 years because there was no other rational explanation for the precession. Now there is! The reason for that assist in acceleration is expanding space.

True Precessions

One may reasonably ask, if Mercury's early arrival due to perturbations, cannot cause orbital precession, then why would an early arrival caused by space expansion cause a precession? Why is an early arrival as a result of space expansion different than planetary perturbations?

Good question! The answer is that the accelerations are applied in different directions. The acceleration due to planetary perturbation is toward the attracting planet – not the Sun. The effect of another planet's gravitational attraction is to draw the planet off course. We call this displacement of the planet's original orbit a perturbation. The perturbed planet's distance from the Sun is slightly changed, as is its velocity. The perturbed planet is sometimes drawn closer to the Sun, sometimes further away. The perturbed planet's path and velocity are constantly being adjusted in the perturbation.

The acceleration due to space expansion is added to the gravitational acceleration from the Sun and is always directed toward the Sun. However, this does not change the planet's mean distance from the Sun, only its velocity. This additional velocity causes the whole orbit to rotate around the focus. This is a true precession! So in order for a planet to experience a precession of its orbit it must have an assist in acceleration. However not all assists in acceleration will cause a precession.

Let's do some simple calculations to show why the assist in acceleration from space expansion will cause a precession. Take a look at the radius of Mercury. A planet's radius from the Sun can be calculated using Equation 39 – Radius of a Planet's Orbit:

Equation 39 – Radius of a Planet's Orbit

$$r = \frac{v^2}{a}$$

Where:

  r   is the radius (semi-major axis) of the planet.
  v   is the planet's velocity
  a   is the planet's acceleration toward the Sun

We'll use the values from Table 23 to calculate the radius of Mercury in two ways. We'll use more decimal places for a precise calculation to demonstrate that the radius does in fact remain the same. In the first calculation we'll be using the normal velocity from the "V1" column in Table 23 – Acceleration of Planets, and the gravitational acceleration from the Sun given in the "Acceleration" column.

$$r = \frac{47878.805668^2}{3.95858 \times 10^{-2}} = 57{,}909{,}150{,}054 \text{ meters}$$

In the second calculation we'll use the velocity of Mercury which includes space expansion. This is obtained from the "V2" column also in Table 23. The gravitational acceleration comes from the "Acceleration" column, but we will add to it the acceleration from expanding space, found in the "X Accel" column.

$$r = \frac{47878.852997^2}{3.95858 \times 10^{-2} + 7.826195 \times 10^{-8}} = 57{,}909{,}150{,}054 \text{ meters}$$

Compare the two calculated radii; notice that they match exactly. This illustrates the point that the assist in acceleration and corresponding velocity increase from space expansion does not affect the orbit in the same way as planetary perturbations. The mean radius is unaffected. The velocity increase does however cause the orbital precession.

This phenomenon also explains how the stars at the outer edge of the galaxy can revolve around the galaxy at high velocities without being thrown off into space. The velocity is increased without an increase of the mean orbital radius. The rotation of galaxies due to space expansion is like a precession.

Precession via Relativity

One may also reasonably ask what makes the relativistic effects of Einstein's theory a precession of Mercury's apsides? This is a totally different phenomena. Einstein pointed out the fact that massive objects affect the space and time in their vicinity. We have seen the affect mass has on space. It's the contraction of space that gives us gravity. Now we see an affect of time distortion on the orbit of Mercury.

Massive objects cause time to flow more slowly. The greater the mass of the object, and the closer we get to it, the slower the flow of time. So as Mercury approaches its perihelion, its closest point to the Sun, time flows slightly slower than at its aphelion. The slower time flows near the perihelion, the faster Mercury will travel around it. This may not seem logical, but consider the following:

Two runners are in a race around a one mile oval track. They are both monitoring their progress with their own clock. The first runner is faster and has a normally running clock. The second runner is slower, but is keeping time with a decelerated clock. The first runner completes the one mile race first with 5 minutes on his clock. The second runner crosses the finish line a minute later, but because his clock was running much slower, it shows only 4 minutes have passed in his trek around the track.

The second runner claims victory by running the mile in less time. As irrational as that sounds, it makes the runner with the slower clock appear to be faster. So it is with Mercury at its perihelion. Time flows slower nearer the perihelion of Mercury's orbit, so Mercury moves faster rounding its perihelion; this results in an orbital precession. And that's why they call it a perihelion precession and not an aphelion precession.

The muddy waters have now been cleared. Below is the revised precession table showing the total of the precessions to add up correctly.

Table 26 – True Precessions

Precession from Space Expansion	532
Precession from effect of Relativity	43
Precession from Perturbation by Planets	0
Observed Precession of Mercury	575

Chapter 21

Circular Orbits

Like a wheel within a wheel.

The Cause of Precession

When doing the analysis of Mercury's orbit to explain how the orbits of planets make the transition from elliptical to circular, it became apparent there is another way to explain the precession of Mercury's orbit. The order of discovery does not necessarily fit into a nice neat chapter where everything on any particular subject is complete. So before proceeding with the explanation of circular orbits, let's take some time to complete the picture of perihelion precession. Some repetition is involved but this section will show cause of precession and will make the precession picture clear.

We have already seen in the preceding chapter, that the acceleration assist given by the expansion of space increases the velocity of the revolving object. The velocity increases, but that does not change the object's mean distance from the center of attraction. A planet under normal gravitational attraction will maintain its angular momentum throughout its whole cycle. It will travel faster as it approaches its perihelion, and slower as it approaches its aphelion. This means that the planet's orbit will maintain its orientation. There may be perturbations, but there will be no precession or recession.

It is the assist in acceleration given by expanding space that causes the orientation of a planet's orbit to become slightly rotated. The difference in the velocity induced by the expansion of space is such that the planet's mean orbital radius is maintained. But because the planet's velocity is increased, a slight rotation of the orbit occurs.

The increase of velocity from expanding space is very small indeed. In the case of Mercury, the average is only 47 millimeters per second. This adds up slowly which is why Mercury's precession is measured in number of arc-seconds per century.

Two Velocities

We now know that there are two different accelerations that determine the motion of the planets. These two accelerations come from two different sources: contracting space and expanding space. Therefore we also have two different velocities. The velocity of a planet that we normally work with using Newton and Kepler's laws is the velocity from the acceleration caused by gravity – space contraction. We understand from those laws that a planet's velocity changes depending on where the planet is on its elliptical orbit. A planet's velocity is faster near the perihelion, and slower near the aphelion.

Table 27 – Mercury's Velocities is very similar to Table 23 – Acceleration of Planets. In this case we are only showing the data for three locations of Mercury's orbit. Row one is for Mercury's aphelion. Row two is for the mean radius of Mercury's orbit, the semi-major axis. Row three is for Mercury's perihelion. The radii of those locations are in the "Radius" column. The gravitational accelerations are shown in the "Acceleration" column. The space expansion accelerations are shown in the "X Accel" column. The "V1" column shows the velocity of Mercury due to the contraction of space, while the "V2" column shows the velocity of Mercury due to the contraction and the expansion of space combined. The "Vx" column shows the velocity of Mercury due to the expansion of space only. The velocity in the "Vx" column was obtained by subtracting the velocity in column "V1" from column "V2".

Table 27 – Mercury's Velocities

Apsides	Radius	Acceleration	X Accel	V1	V2	Vx
Aphelion	6.9817E+10	2.7234E-02	6.4913E-08	43,604.99	43,605.04	0.05197
Mean	5.7909E+10	3.9586E-02	7.8262E-08	47,878.81	47,878.85	0.04733
Perihelion	4.6001E+10	6.2733E-02	9.8520E-08	53,719.52	53,719.56	0.04218

As was already mentioned, a planet's velocity is faster near the perihelion, and slower near the aphelion. This can be seen from the normal velocities in the "V1" column due to space contraction, and can be derived from Kepler's laws. The velocity of Mercury at its aphelion is about 43.6 km/s. The velocity of Mercury at its perihelion is about 53.7 km/s. Mercury travels about 10.1 km/s faster at its perihelion than at its aphelion.

Now let's examine the space expansion velocities in the "Vx" column. The velocity from space expansion is greater at the aphelion than at the perihelion by just .00979 m/s. The values come from the "Vx" column of Table 27, and the calculation is shown below:

$$0.05197 - 0.04218 = .00979 \text{ m/s}$$

Or let's just say that is about 1 centimeter a second faster at the aphelion than at the perihelion. This is just the opposite of the velocities from space contraction. This has a very small but definite effect on the motion of planets, as we shall soon see.

The velocity we had not previously known about is that which is induced by expanding space. So, of course, no one had ever used it in any calculation for the planetary orbits. But now we can see that this extra velocity is the real cause of Mercury's perihelion precession, as shown in Figure 43.

The Fundamental Force

Figure 43 – Precession of Mercury's Orbit

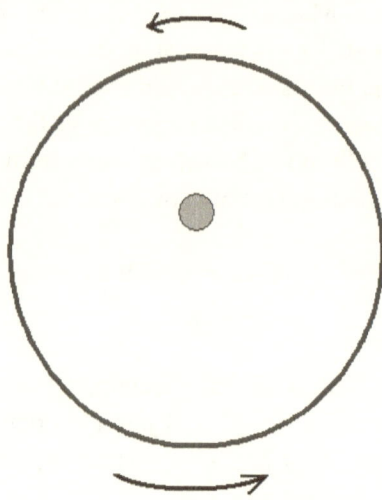

Figure 43 – Precession of Mercury's Orbit illustrates how the precession of Mercury's orbit occurs as a result of the increased velocity due to space expansion. The increased velocity due to space expansion is slightly smaller at the perihelion (top of figure) than at the aphelion (bottom of figure); as depicted by the length of the arrows. This difference of velocity causes a slight rotation of Mercury's orbit around the main focus. This is the cause of the precession of Mercury's orbit.

Orbits in Transition

One of the great mysteries of science still waiting to be solved is the question of why the planets revolve around the Sun in nearly circular orbits. Newtonian dynamics show that once a planet, or the debris composing a planet, is captured by the Sun, it will establish an elliptical orbit. There is nothing in Newton or Kepler's laws to explain how these orbits eventually become circular. The hypothesis presented here is that the circular motion of the moons, planets and galaxies is the result of expanding space.

We've seen in the previous section that there are two different velocities due to two different accelerations. In order to truly understand the motion of planets, both velocities must be accounted for. Because space expands in all directions at the same rate, the orbits it induces are perfectly circular. The direction for the velocity from expanding space is perpendicular to the radius. Also, the velocity from expanding space is faster at the aphelion than the perihelion. This means that there would be a tendency for the orbital path around the aphelion to be shortened, and the orbital path around the perihelion to be increased. This is illustrated in Figure 44 – Orbits of Mercury. Also remember that the acceleration and resulting velocity of objects due to space expansion do not follow Kepler's Laws. Kepler's Laws only apply to space contraction.

Figure 44 shows two possible orbits for the planet Mercury. The lower ellipse represents Mercury's current elliptical orbit. The upper circle represents the circular orbit that Mercury will eventually have. The Sun is at the focus of both orbits. The mean radius of both orbits is the same. The arrows indicate the change in the direction of the elliptical path caused by the additional velocity from expanding space. The arrows are highly exaggerated, but they illustrate the direction of the additional velocity which is parallel to the circular orbit, and perpendicular to the orbit's radius. This additional velocity would have a tendency to push the perihelion out a little further from the Sun while drawing the aphelion in a little closer. Notice that the arrows are located outside the perihelion and inside the aphelion. Later in this chapter we will see the transition Mercury is making from an elliptical to a circular orbit.

Figure 44 – Orbits of Mercury

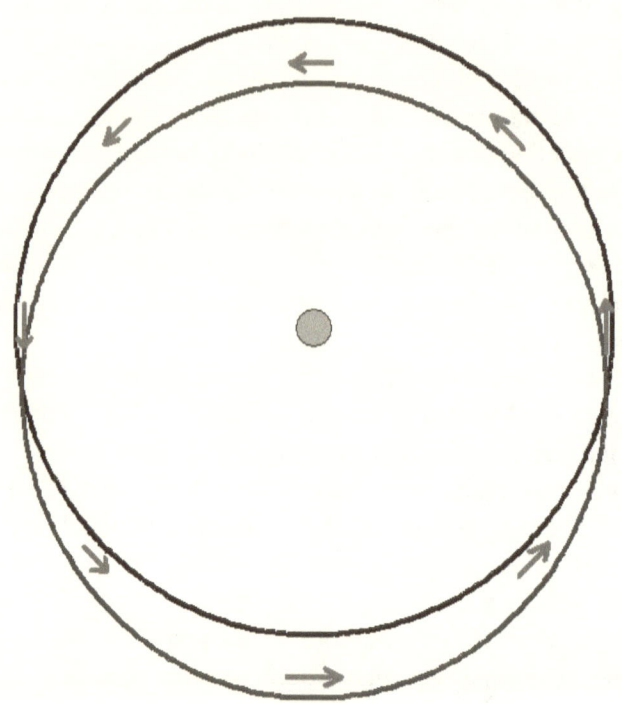

Change of Focus

We have already shown that the increase of velocity from space expansion does not affect the mean radius of the planet's orbit. The mean radius for a planet is the same regardless of whether it has a circular or an elliptical orbit. However, the radius at the perihelion and aphelion do change as described in the previous section. This has a very interesting effect on the dynamics of a planet's orbit. Instead of a reduction in the size of the orbit's mean radius there is a reduction in the distance from the center of the ellipsis to the main focus. The center of the ellipse gradually moves toward the focus at which the Sun is located. The method of calculation to obtain the reduced offset from the center of the ellipse to the focus is given in the following equation:

Equation 40 – Offset of Focus from Center

$$o = \frac{o_c}{1 + ta_x^2}$$

Where:

    o is the offset of the focus from center

    o_c is the current offset of the focus from center

    a_x is the acceleration from expanding space

    t is the period (time of orbit in seconds)

The equation to calculate the acceleration from space expansion was already given in the Space Expansion chapter. The space expansion acceleration for Mercury was also given in the "X Accel" column of Table 27 – Mercury's Velocities, and is 7.8262×10^{-8} m/s^2. The current offset of Mercury's focus from center is 11,907,903,751 meters. The time frame we'll use is one complete cycle of Mercury around the Sun, which takes 7,600,533 seconds. Let's calculate the new offset after one complete cycle of orbit:

$$o = \frac{11,907,903,751}{1 + (7,600,533 * (7.8262 \times 10^{-8})^2)} = 11,907,903,197 \text{ meters}$$

The new offset of Mercury's focus from center is 11,907,903,197 meters. The difference between the original offset and the one just calculated is 554 meters. This means that with every cycle of Mercury's orbit the offset of the center of the elliptical orbit to the focus is decreased by 554 meters. This is only .0729 millimeters every second, but it adds up. Also decreased by 554 meters every time Mercury orbits the Sun is the aphelion of Mercury's orbit. If astronomers could accurately measure the decrease in Mercury's aphelion it could possibly validate the expansion of space. If no decrease is found, the claim that space expansion is the cause of circular orbits would be invalidated.

Note that these calculations are only approximations. More accurate numbers can only be obtained through the use of calculus, which is beyond the scope of this book. However, these equations and calculations are close enough to illustrate the concepts being presented.

Change of Eccentricity

Now we can calculate the new eccentricity of Mercury's orbit by dividing the offset just calculated by the average distance from the Sun (semi-major axis). This is shown in the equation below:

Equation 41 – Eccentricity

$$e = \frac{o}{r}$$

Where:

- e   is the eccentricity of the planet's orbit
- o   is the offset of the focus from center
- r   is the average distance from the Sun (semi-major axis)

Let's do the calculation for the eccentricity of Mercury's orbit.

$$e = \frac{11{,}907{,}903{,}197}{57{,}909{,}175{,}000} = 0.20563068$$

The eccentricity of Mercury's orbit before the cycle began was 0.20563069. The new eccentricity calculated after Mercury made a complete cycle in its orbit is 0.20563068. This is a difference of only 0.00000001. This is a very small decrease of eccentricity. But it is not insignificant! The eccentricity of Mercury's orbit decreases with each revolution around the Sun. As previously stated, it may not be much, but it adds up.

Let's make a quick calculation to get the approximate amount of the time required to reduce the eccentricity of Mercury's orbit by 10 percent.

We can do that by taking 10 percent of the focus's offset and dividing that by the amount of reduction each cycle. This will give us the number of times Mercury must cycle around the Sun to reduce its orbital eccentricity by 10 percent.

$$\text{number of cycles} = \frac{.10 * 11,907,903,197}{554} = 2,149,441$$

The number of times Mercury needs to cycle the Sun is 2,149,441. This number can be converted into Earth years by multiplying it by Mercury's orbital period in years. Mercury's orbital period is about .2408467 years. That times the number of cycles is 517,687 Earth years. This is the approximate number of years required to reduce the eccentricity of Mercury's orbit by 10 percent. This is only a gross approximation because the offset reduction is itself reduced with every cycle. An accurate solution to this problem requires a calculus solution.

However, the point can still be made. But what is the significance of all this? This is very significant! It means that a planet's eccentricity is reduced with each cycle. This applies not only to Mercury, but to all planets and moons everywhere. This means all elliptical orbits will slowly shift toward more circular orbits. This explains why the planets and moons have very nearly circular orbits. The inducement toward circular orbits is the assist in acceleration which gives the extra orbital velocity. This can all be attributed to the expansion of space!

Even though Mercury's eccentricity would be reduced by 10 percent in one half million years, it would still be more eccentric than the other planet's in the solar system as they currently are. The transition from elliptical orbits to very nearly circular orbits would take millions of years.

Visualizing the Transition

Figure 45 – Mercury's Orbit in Transition, shows the transition that Mercury's orbit would make as a result of the dynamics just discussed. These dynamics describe how planets transition from elliptical to circular orbits. As in Figure 44 this figure shows two orbits superimposed on each other. The lower ellipse represents Mercury's elliptical orbit as it currently exists. The upper circle represents the circular orbit Mercury will eventually have. The disk in the middle represents the Sun, which is at the focus of the elliptical orbit, and it's also at the center of the circular orbit. The size of the Sun is highly exaggerated. However, the orbits are in proper proportion to each other.

Figure 45 – Mercury's Orbit in Transition

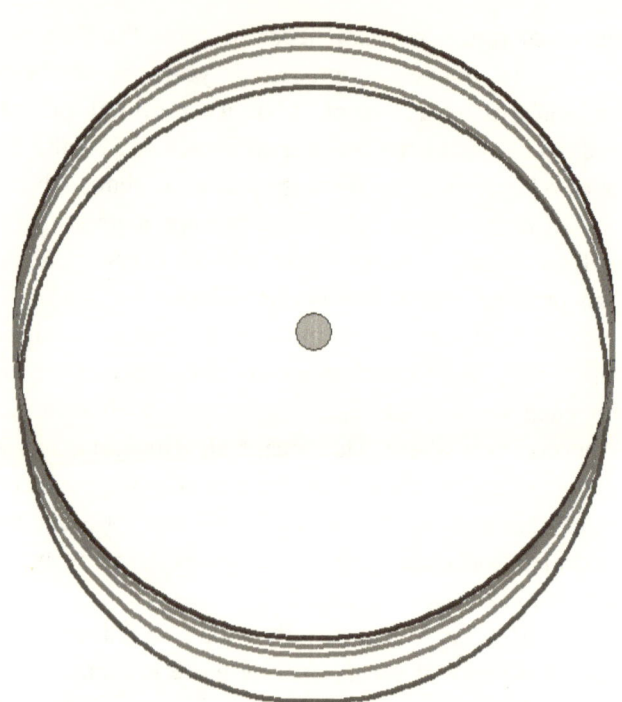

The gray line in this diagram shows the transition from one orbit to another. The path of the transition is highly exaggerated. It shows the transition being made in three cycles, whereas in reality, the transition would take 2,149,441 cycles. With each cycle the orbital path becomes more circular. Although highly exaggerated, it does visually illustrate the idea.

In this transition Mercury starts at the right and rotates around the Sun in a counter clockwise direction. The first cycle shows Mercury's path at the lowest level closest to the elliptical orbit. In the second cycle Mercury take the orbital path in the middle. In the third cycle Mercury is in the highest orbital path closest to the black circular orbit. In time the orbital path of Mercury will become circular.

This transition takes place because of expanding space. Each second the center of Mercury's elliptical orbit moves .0729 millimeters closer to the focus. After each cycle it has moved 554 meters closer to the focus. After 2,149,441 cycles around the Sun the eccentricity of Mercury's orbit is reduced by 10 percent. But even a 10 percent reduction in Mercury's eccentricity will not make it as circular as the other planets. One last reminder here is that this is a gross approximation of the time required to make this transition. After each cycle the amount of distance that is reduced from the center of the ellipsis to the focus shrinks. In simple terms, the 554 meters gets smaller and smaller as Mercury's orbital eccentricity is reduced. So the approach gets shorter and shorter. This can be seen in Figure 45 – Mercury's Orbit in Transition. The orbital paths get closer and closer to each other as they approach a circular orbit.

Space Expansion Observations

We have seen three different phenomenon that effect the period of a planet's orbit. The first of these is perturbations caused by the proximity of other planets. These perturbations can alter the planet's orbital period by inducing it to travel faster or slower, depending on whether the attracting planet is in an inferior or superior orbit. During these perturbations a planet's speed and position are influenced according to another planet's attraction. This adjustment does not cause a precession.

The second phenomena is the effect of relativity (space contraction). In this case there is a time and space distortion near the focus of the Sun which increases the planet's speed traveling around the perihelion. In this case the planet makes no adjustment in its mean radius; but the period of revolution is shortened resulting in a precession of the perihelion.

The third phenomenon is space expansion. In this case the planet's mean radius remains the same, but the overall period of revolution is shortened, resulting in a precession. Also with each revolution there an adjustment to the offset of the center of the ellipse to the focus resulting in a more circular orbit.

The expansion of space is a very real phenomenon. The affects of space expansion are real laws of physics that we are only now beginning to understand. The evidence is accumulating. The phenomenon of space expansion can logically explain a variety of other phenomena. Let's briefly look at the following examples:

1. The Pioneer anomaly and the orbits of other spacecraft can be more accurately predicted and determined by adding space expansion acceleration to the gravitational acceleration.
2. It has now become clear why there are high tides on both sides of the Earth. The combined effects of gravity and space expansion will restructure the space around the planet to cause the tides. This also explains why the tidal effect diminishes with the inverse of the distance cubed.
3. The expansion of space explains how galaxies rotate and why stars on the outer rim revolve so fast, as well as why spiral arms still exist over billions of years.
4. Space expansion effectively explains what cosmologists are currently attributing to dark matter. It also explains the apparent effects of gravitational increase over distance that is described by the MOND theory.
5. Orbital precession is now explained.
6. Finally there is an answer to the question of why the planets and their moons have nearly circular orbits.

Once again, these are all effects of space expansion.

Chapter 22

The Fate of the Universe

Is that all there is to the universe?

Future Expectations

Currently scientists speculate that the universe began with a "Big Bang". That would be where all the matter in the universe explodes out of a big black hole called a singularity. All the mass would spread out throughout the universe forming stars galaxies and whatever else is contained in the universe. But what happens after that? Ever since Hubble discovered that the universe is expanding scientists have speculated on its fate. What will eventually happen? Thus far, only three possibilities have been presented. A brief description of each follows.

The first scenario goes back to Newton's time. Newton realized in his time that eventually the force of gravity would attract all the matter in the universe together. Gravity would slow down, and ultimately halt, the expansion of the universe, at which time the expanding universe would become the contracting universe. All the mass contained within would eventually come together in one gigantic crush, often called "The Big Crunch." Everything in the universe would be sucked into one super black hole – called a singularity. Some scientists speculate that this could be a process that happens over and over again. The universe comes into existence through a big bang and then collapses back onto itself in a big crunch, and the cycle repeats itself.

A second possible fate of the universe is that it will remain in a stable state forever. This could only happen if the forces causing the expansion of the universe were perfectly balanced with the contracting forces. Einstein's field equations do not permit a perfectly balanced universe. It must be either expanding or contracting. Willem De Sitter's mathematical analysis of Einstein's field equations showed it to be expanding. Even if it were perfectly balanced, time and entropy would have their way. Over time stars would consume their fuel and burn out. Eventually all the stars that could have formed would have done so, then burnt out. All the energy would be consumed, all interactions of mass and energy would have completed, and everything would have fallen to the lowest common denominator. The ultimate end would be a cold, dead universe.

The third possible fate of the universe ends much the same. The only difference being that the universe expands forever. Galaxies will grow farther and farther apart, eventually disappearing from view, which wouldn't matter anyway because all the stars in all the galaxies would eventually burn out. With all fuel exhausted there would be no light emitted from anywhere. All galaxies would become cold and lifeless.[32]

None of these scenarios are very appealing. But those are the scenarios we have. The latest scientific information presented by Saul Perlmutter and his team on the Supernova Cosmological Project shows that the universe is expanding at an ever faster rate. It would appear that the third scenario presented is the fate to which we must resign.

The ultimate death of the universe is demoralizing. Even though the lifespan of the universe may be a trillion years, it all seems quite pointless. Why would the universe come into existence only to die a cold lonely death? This of course is the same problem we face as humans. What is the point of life if death is the final reward? Many of us search for things to give meaning to our lives. Some look to religion, others look within themselves, and some find it in a great cause. In any case, we're all in the same boat, and facing the same end.

If the universe dies, so does mankind. So does all life on all planets everywhere in the universe. But there is another alternative for the fate of the universe. This new view will change the way we see the universe. It may also change the way we view our life!

The three scenarios described above, and only these three, are what has been presented to us. There doesn't seem to be any other reasonable alternatives. Why only these three? Because these are the scenarios based on all the known laws of physics. There are forces, effects, expansion, contraction, mass and motion, etc. Everything must be considered. The laws of physics must be obeyed, and the results are before us. Before any new possibilities can be presented for the fate of the universe there must be a good understanding of how these laws work.

What will be presented in subsequent chapters is a totally different scenario for the fate of the universe. This scenario may seem amazing, but is neither fantasy nor dream; it is based on solid scientific knowledge and pure simple logic. We're going to look at the universe in a different way. Once again, we may have to discard prior beliefs that don't hold up under scrutiny; we will have to see things in a new light – as they really are. Before we do, let's understand the basis of the Big Bang and Hubble's Law.

Basis for the Bing Bang

In 1927 Belgian priest, physicist, and astronomer Georges-Henri Lemaitre (1894-1966) presented a theory explaining the creation of the universe. His theory proposed that the universe came from one single source; he called it "The Hypothesis of the Primeval Atom."[33] Later that same hypothesis became known as The Big Bang Theory, and Lemaitre became known as "The Father of the Big Bang."

One of the problems with the big bang theory is that it always begins with an incredible amount of matter contained in a minuscule space called a singularity. This space, or place, is even more highly condensed than any super massive black hole, and cannot be described by the current laws of physics. Then suddenly, for unknown reasons, all the matter that exists in the universe explodes out of this singularity. This moment in time is sometimes described as "the initial awkward moment" since it cannot be justified in scientific terms. As things begin to cool elementary particles are formed, coalescing into galaxies and the stars within.

The problem is where all this mass originally came from? This has been a question asked repeatedly by generations of students of science and religion. In presenting their case, proponents of the Big Bang Theory have abided by the laws of physics as best they can. The law of conservation of mass is strictly adhered to. This law states that the mass in a particular frame of reference will remain constant. When discussing a frame of reference we mean in a closed frame where nothing from outside the frame comes in and nothing gets out. Within this closed frame of reference the amount of mass in existence will remain the same. Regardless of what chemical reactions take place, including fire, the amount of matter in the frame of reference must remain the same.

The law essentially says that matter cannot be created or destroyed. It can change forms: solid, liquid, or gas; or be rearranged in molecular structure, but the amount of mass will always remain constant. So if our frame of reference is the universe, where nothing gets in and nothing gets out, the universe is closed. All the matter that ever existed still exists and will exist throughout eternity. The first to clearly identify and document this law was French chemist Antoine Laurent Lavoisier (1743-1794).

There is still another law that is analogous to this: the law of conservation of energy. It states that all the energy in a closed system will remain constant. Energy can be converted to different forms; for example, light energy can be converted to thermal energy but the overall amount of energy in a closed frame of reference will remain the same. These laws essentially state that energy cannot be created or destroyed, it can only be changed from one form to another. Once again, all the energy that exists in our frame of reference, the universe, will remain the same throughout eternity.

However, Einstein showed us that matter and energy are the same thing in different forms. Matter can be converted to energy, and energy into mass. So then these laws of conservation need to be combined. Now we would have to say that the mass and energy in a closed frame of reference will remain the same no matter what happens within. Even in a nuclear explosion mass would be converted to energy, but the total amount of mass and energy combined would still be the same.

Considering these laws, we readily see why the big bang is thought to be the only reasonable way the universe could have been created. But we still have no idea where the matter came from in the first place. Theorists would say that the matter and energy have always existed. However, even the big bang would have to violate the law of conservation of energy. Where did a sudden burst of energy come from to cause the big bang? Somehow a vast amount of energy would have to be injected into the universe to cause all the matter to be scattered throughout.

It seems that the big bang leaves us with few options and no alternatives. Any theory proposed that starts with no matter in existence in the universe must of necessity violate the laws of conservation. In the past we have witnessed events that appear to violate known laws. Upon further study and investigation we find another law that supersedes the first. So here we need further study and investigation. But first let's understand Hubble's Law.

Chapter 23

Hubble's Law

Hubble bubble toil and trouble.

Hubble's Constant

In 1999 an international group of astronomers called the Hubble Space Telescope Project Team completed an eight year effort to measure the expansion of the universe.[34] This was done in an attempt to determine a more precise rate of expansion than what was currently known. The team found almost 800 Cepheid variables in the 18 galaxies observed, which provided the means for a more accurate measurement. The expansion rate of the universe was found to be about 70 kilometers per second per million parsecs (Mpc), with an uncertainty of 10 percent. This rate is called the Hubble constant.

In order to understand the significance of that number, we must first understand Hubble's Law. During his life Hubble discovered that the galaxies are growing farther apart. There were exceptions, of course. Galaxies in close proximity are attracted to each other; they are on a collision course, or have already collided. Galaxies in collision form larger galaxies. Galaxies with sufficient distance between them will grow more distant over time. This trend became known as Hubble's Law. Simply put, the law states that the greater the distance between galaxies, the faster they move away from each other. After many years of observation that trend held true, and so became a law.

The Hubble constant then tells us the rate at which galaxies recede from each other. For every million parsecs (Mpc) between galaxies, they recede at about 70 kilometers per second. This number is believed to be accurate to within 10%. One parsec is about 3.08609×10^{16} meters. One million parsecs is 3.08609×10^{22} meters. To put it in perspective, our galaxy the Milky Way, is only about $1/30^{th}$ of one Mpc across.

Even more recently other teams of astronomers using other methods of measurement have produced similar results. A team at NASA's Chandra X-ray Observatory, came in at 77 km/sec/Mpc, but with an uncertainty of ± 15%.[35] We may not know the exact number, but the trend is real and we have a good approximation.

Using Hubble's Law and Hubble's constant we can readily calculate the age of the universe. All we need do, to look back in time, is to reverse the trend, and calculate the time it takes for all the galaxies to collide in one big crunch. Then we could say that was the amount of time that has passed since the big bang. Thus, Hubble's Law is often presented as evidence that there must have been a big bang somewhere back in time. Let's use Hubble's constant now to make that calculation. By dividing the distance between the galaxies by the rate of expansion we can calculate the time required to traverse that distance.

Distance Between:	3.08609E+22	meters
Expansion Rate:	70,000	meters/s
	4.4087E+17	seconds

The time required to expand that distance at the given rate is $4.4087e^{17}$ seconds. That converts to 13.97 billion years. This is why many scientists believe the universe is about 14 billion years old. That would be the time required for the universe to expand from a big bang to its current state. Had we done the same calculation for 77 km per second per Mpc, we would have calculated the age of the universe to be about 12.7 billion years old.

The Rate of Expansion

Table 28 shows the rate of expansion according to Hubble's Law. The column on the left, "Mpc" is the distance in millions of parsecs. The table ends at 4,282 million parsecs. This is believed to be the radius of the universe. If we multiplied 4,282 million parsecs by 3.262 to get the equivalent in light years, then multiplied again by the speed of light, 299,792,458 meters per second, then multiply once again by the number of seconds in a year, (3.15576×10^7), we would get the approximate radius of the universe in meters – 1.32146×10^{26} meters.

Table 28 – Hubble's Law

Mpc	M Light Years	Velocity	Relativistic	Expansion	Difference
100	326.20	7,000,000	6,998,092	70,000	1,908
200	652.40	14,000,000	13,984,726	69,866	15,274
300	978.60	21,000,000	20,948,415	69,637	51,585
400	1,304.80	28,000,000	27,877,608	69,292	122,392
500	1,631.00	35,000,000	34,760,657	68,830	239,343
600	1,957.20	42,000,000	41,585,787	68,251	414,213
700	2,283.40	49,000,000	48,341,059	67,553	658,941
800	2,609.60	56,000,000	55,014,330	66,733	985,670
900	2,935.80	63,000,000	61,593,219	65,789	1,406,781
1000	3,262.00	70,000,000	68,065,062	64,718	1,934,938
1100	3,588.20	77,000,000	74,416,864	63,518	2,583,136
1200	3,914.40	84,000,000	80,635,249	62,184	3,364,751
1300	4,240.60	91,000,000	86,706,405	60,712	4,293,595
1400	4,566.80	98,000,000	92,616,019	59,096	5,383,981
1500	4,893.00	105,000,000	98,349,209	57,332	6,650,791
1600	5,219.20	112,000,000	103,890,440	55,412	8,109,560
1700	5,545.40	119,000,000	109,223,437	53,330	9,776,563
1800	5,871.60	126,000,000	114,331,075	51,076	11,668,925
1900	6,197.80	133,000,000	119,195,260	48,642	13,804,740
2000	6,524.00	140,000,000	123,796,783	46,015	16,203,217
2100	6,850.20	147,000,000	128,115,155	43,184	18,884,845
2200	7,176.40	154,000,000	132,128,402	40,132	21,871,598
2300	7,502.60	161,000,000	135,812,827	36,844	25,187,173
2400	7,828.80	168,000,000	139,142,713	33,299	28,857,287
2500	8,155.00	175,000,000	142,089,970	29,473	32,910,030
2600	8,481.20	182,000,000	144,623,686	25,337	37,376,314
2700	8,807.40	189,000,000	146,709,571	20,859	42,290,429
2800	9,133.60	196,000,000	148,309,245	15,997	47,690,755
2900	9,459.80	203,000,000	149,379,318	10,701	53,620,682
3000	9,786.00	210,000,000	149,870,179	4,909	60,129,821
3100	10,112.20	217,000,000	149,724,373	-1,458	67,275,627
3200	10,438.40	224,000,000	148,874,368	-8,500	75,125,632
3300	10,764.60	231,000,000	147,239,408	-16,350	83,760,592
3400	11,090.80	238,000,000	144,720,930	-25,185	93,279,070
3500	11,417.00	245,000,000	141,195,650	-35,253	103,804,350
3600	11,743.20	252,000,000	136,504,636	-46,910	115,495,364
3700	12,069.40	259,000,000	130,435,037	-60,696	128,564,963
3800	12,395.60	266,000,000	122,687,242	-77,478	143,312,758
3900	12,721.80	273,000,000	112,809,810	-98,774	160,190,190
4000	13,048.00	280,000,000	100,051,470	-127,583	179,948,530
4100	13,374.20	287,000,000	82,943,137	-171,083	204,056,863
4200	13,700.40	294,000,000	57,514,244	-254,289	236,485,756
4282	13,967.88	299,740,000	5,607,073	-519,072	294,132,927

The second column, "M Light Years", shows the same distance in millions of light years. Notice the last row ends with 13,967 million years, better expressed as 13.96 billion years. This is what is believed to be the age of the universe.

Now notice the third column, "Velocity." The numbers in this column have been determined by the Hubble constant for the recession of galaxies. The velocity in each row of this column is 7,000,000 meters (7,000 km) greater than the previous. Each row is 100 million parsecs, so the rate of expansion remains the same: 70 km per second per million parsecs. We see here the increasing speed at which galaxies recede from each other with each 100 mega parsecs. That is 7 million meters per 100 million parsecs, which is 70 km per second per million parsecs. Once again, the table ends at 4,282 million parsecs where the speed of recession approaches the speed of light.

The "Relativistic" column shows the relativistic velocity of receding galaxies based on the Lorentz transformation. In Einstein's book on Relativity there is a chapter entitled "The Lorentz Transformation" that explains what the transformation is all about. Briefly stated here, we may say it takes into account the effects of relativity when determining the speed of an object. The equation for the Lorentz transformation is shown below:

$$Rs \ = \ 70 \text{ km} * \text{Mpc} * \ \sqrt{1 - \frac{(70 \text{ km} * \text{Mpc})^2}{c^2}}$$

The "Difference" column is the difference between velocity and the relativistic velocity. The "Expansion" column shows the Hubble constant for each 100 million parsecs based on the relativistic speeds from the "Relativistic" column. The rate of expansion in the "Expansion" column shows the effects of relativity as speeds increase. Notice that the relativistic expansion rates start to decline very slowly for each 100 million parsecs. The items in this column were calculated by subtracting the relativistic velocity from the previous relativistic velocity and dividing by 100. This gives the average relativistic velocity for that mega parsec.

Figure 46 is a graphical representation of Table 28. The black line illustrates the Hubble constant from the "Velocity" column. The gray line is the calculated relativistic expansion rate resulting from the Lorentz transformation.

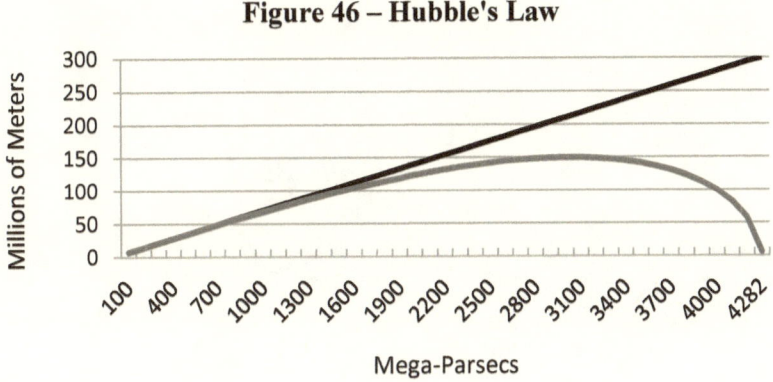

Figure 46 – Hubble's Law

We see galaxies moving away from us at 70 km/sec/Mpc. That appears to be Hubble's constant for the first 1,600 million parsecs, which is equivalent in time to about 5.2 billion years. For about the first 5.2 billion years Hubble's constant, the expansion rate of the universe, appears to be constant. The relativistic expansion rate and the Hubble constant are very close. Beyond that the expansion rates begin to diverge.

At 2,200 million parsecs the rate of relativistic expansion is only about 40 km/sec/Mpc. That value comes from the "Expansion" column in Table 28 – Hubble's Law. It appears that 7 billion years ago the universe was expanding only 40 km/sec/Mpc according to the relativistic time. At 2,800 million parsecs the rate of expansion appears to be only about 16 km/sec/Mpc. And so it appears that 9 billion years ago the universe was expanding at only 16 km/sec/Mpc. By taking the effects of relativity into consideration it would appear that the universe is expanding faster now than it did in the past. This suggests an ever increasing rate of expansion.

What has been discussed in this chapter is the rate of expansion of the universe according to Hubble's Law. Seeing the galaxies moving away from each other gives us the impression that the galaxies are carried along with the expansion. This is not the case as we shall see in the next chapter.

Chapter 24

Galactic Free Fall

All galaxies are falling.

Galactic Motion

In this chapter we will see that according to Hubble's law all the galaxies are in free fall. In order to get a better grasp of galactic motion we must first analyze the data we've accumulated so far. Then we'll graph the related information to enhance our understanding of what is really happening. In this chapter we will use Hubble's constant to show that the galaxies are falling.

Let's look at Table 29, which is only a partial listing of Table 28, but enough to illustrate the motion of galaxies. From it we will acquire some very interesting information about galactic motion. As with Table 28 – Hubble's Law, the first column in this table (Mpc) is the span of distance in millions of parsecs. The values in the "Mpc" column increase by 100 Mpc in each subsequent row.

The "Time in Seconds" column is the number of seconds necessary for space to expand to the size specified in the "Mpc" column. Each subsequent row increases by 1.0294×10^{16} seconds.

The "Rate of Acceleration" column shows the rate at which galaxies are receding from each other. This rate is the same throughout the universe.

The "Velocity" is the Hubble constant (70 km/s/Mpc). This is the velocity of galaxies in motion at the end of each 100 Mpcs. In Table 29 we'll write it as 7,000,000 meters per 100 million parsecs. Each subsequent row increases by 7,000,000 m/s.

The "Distance each 100 Mpc" column shows the distance a galaxy travels in meters in the specified Mpc. Each subsequent row increases by $(7,000,000 * 1.0294 \times 10^{16} = 7.2058 \times 10^{22}$ meters).

The "Total Distance" column shows the total distance traveled by galaxies in their voyage through the mega parsecs.

Table 29 – Galactic Free Fall

Mpc	Time in Seconds	Rate of Acceleration	Velocity	Distance each 100 Mpc	Total Distance
100	1.0294E+16	6.8000E-10	7,000,000	3.6029E+22	3.6029E+22
200	2.0588E+16	6.8000E-10	14,000,000	1.0809E+23	1.4412E+23
300	3.0882E+16	6.8000E-10	21,000,000	1.8015E+23	3.2426E+23
400	4.1176E+16	6.8000E-10	28,000,000	2.5221E+23	5.7647E+23
500	5.1470E+16	6.8000E-10	35,000,000	3.2426E+23	9.0073E+23
600	6.1765E+16	6.8000E-10	42,000,000	3.9632E+23	1.2971E+24
700	7.2059E+16	6.8000E-10	49,000,000	4.6838E+23	1.7654E+24
800	8.2353E+16	6.8000E-10	56,000,000	5.4044E+23	2.3059E+24
900	9.2647E+16	6.8000E-10	63,000,000	6.1250E+23	2.9184E+24
1,000	1.0294E+17	6.8000E-10	70,000,000	6.8456E+23	3.6029E+24
1,100	1.1323E+17	6.8000E-10	77,000,000	7.5662E+23	4.3595E+24
1,200	1.2353E+17	6.8000E-10	84,000,000	8.2867E+23	5.1882E+24
1,300	1.3382E+17	6.8000E-10	91,000,000	9.0073E+23	6.0890E+24
1,400	1.4412E+17	6.8000E-10	98,000,000	9.7279E+23	7.0617E+24
1,500	1.5441E+17	6.8000E-10	105,000,000	1.0449E+24	8.1066E+24
1,600	1.6471E+17	6.8000E-10	112,000,000	1.1169E+24	9.2235E+24

What has just been described in this table is very similar to Table 6 – Free Fall in the chapter The Flow of Space. Table 29 describes the properties of galactic motion and how that motion conforms to all the requirements of a free falling object. Except for the rate of acceleration there is no difference between the fall of a ball dropped off the Leaning Tower of Pisa by Galileo and the free falling galaxies in our universe. All galaxies are falling!

Look at Figure 47 which is a graphic representation of Table 29. Compare it with Figure 4, which is a graphic representation of free falling objects. The curves are similar. The galaxies are falling just like the balls dropped from the Leaning Tower. The acceleration rate and the time frames may be different, but all galaxies are in free fall just the same.

From our perspective as observers from somewhere near the center of the universe, we see galaxies moving away from us toward the outer edge. There is no mysterious force being applied to the galaxies to cause this phenomenon; they are simply falling, just like everything else. The cause of this free fall is the expansion of space. Expanding space adds acceleration to all bodies in the universe. But from what we learned earlier we know that this acceleration is in the direction *opposite* of expansion, which should be inward. How can this be explained when we observe galaxies moving away from us?

Figure 47 – Galactic Free Fall

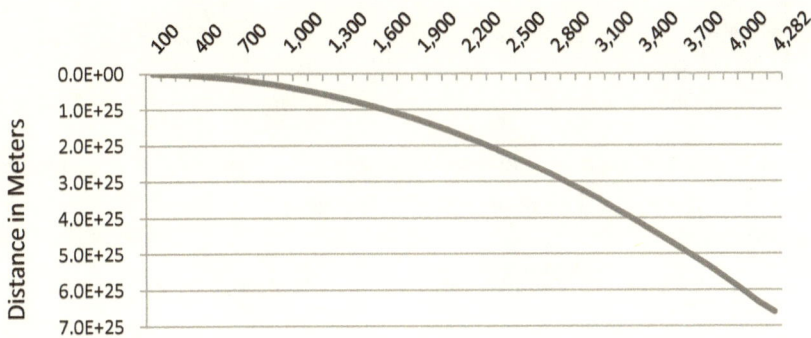

Time in Mega Parsecs

We saw the rate of acceleration in Table 29 – Galactic Free Fall, but we didn't show how we acquired it. We can easily calculate the acceleration caused by expanding space. Acceleration is equal to velocity divided by time. We've got the velocity and we've got the time. The velocity is in the "Velocity" column and the time is in the "Time in Seconds" column. Now let's calculate the acceleration rate of objects in space resulting from the expansion of space. In this case we'll use the data in the first row of Table 29 – Galactic Free Fall.

$$a = \frac{v}{t} = \frac{7,000,000}{1.0294 \times 10^{16}} = 6.8000 \times 10^{-10} \text{ m/sec}^2$$

The acceleration of galaxies is 6.8000×10^{-10} meters per second squared. This number could have been calculated from the data in any row of the table. The acceleration rate is constant across the expanse of the universe. The rubber band illustration applies. Space is expanding across the radius of the universe like a rubber band stretches evenly. We can check the validity of the acceleration by using the equations for free falling bodies. The distance a body falls is equal to the acceleration times the time squared divided by 2. See Equation 42:

Equation 42 – Distance of Free Fall

$$d = \frac{at^2}{2}$$

Let's use the data in the first row again to calculate the distance a galaxy falls in its first 100 mega-parsecs of space travel.

$$d = \frac{at^2}{2} = \frac{6.8000 \times 10^{-10} * (1.0294 \times 10^{16})^2}{2} = 3.6029 \times 10^{22}$$

There is another common equation that gives us the distance traveled. That is: distance is equal to average velocity times time. Let's check by multiplying the velocity of the galaxy by the time traveled, then divided by 2 for the average; once again using the data from the first row:

$$d = \frac{vt}{2} = \frac{7,000,000 * 1.0294 \times 10^{16}}{2} = 3.6029 \times 10^{22}$$

These numbers match and conform to all astronomical observations and the rules of free fall. There can be no doubt – all galaxies are falling!

The Trouble with Hubble

All we really know from Hubble's Law is that galaxies are receding at a relatively constant rate. But now we understand that galaxies are falling, a seeming contradiction. Let's look at Hubble's expansion rate from two different perspectives. First let's assume we are observers located near the center of the universe. From this position we see galaxies moving away from us at about 70 km/sec/Mpc toward the edge of the universe. The farther from us they are the faster their velocity. It's as if they were trying to fall out of the universe.

We've assumed that galaxies are moving outward with the expansion of space; and yes, they are moving outward, but not with the flow of space. Remember in the chapter on Galactic Rotation we saw that stars in a galaxy do not move along with the expansion of space; they are accelerated in the direction opposite the direction of space expansion. The same is true of all the galaxies in the universe.

As space expands outward from the center of the universe, galaxies are accelerated in the opposite direction – toward the center. This makes more sense because we already know that everything falls down – not up. When considering the universe we need to know the difference between up and down. When we speak of falling down on the Earth, we mean down toward the center of the Earth. Going up would be away from center. We speak of the moon falling toward the Earth, and the Earth falling toward the Sun, and the Sun likewise is falling toward the center of the galaxy. Objects fall down, not up. So it seems quite logical that galaxies fall toward the center of the universe. When considering the universe, up is outward toward the rim, and down is toward the center.

We said we were going to view Hubble's expansion rate from two different perspectives. The second perspective is at the outer rim of the universe moving along with the expansion, we would still see galaxies accelerating away from us. The further away the galaxy, the faster its rate of recession. But they would be falling toward the center – falling down. The free fall curve in Figure 47 still applies, only now we're able to view the fall as being in the proper direction. From our perspective at the outer rim we would only see galaxies on one side of ourselves, all of them falling toward the center of the universe. On our other side, looking outward, we would see nothing – because there are no galaxies beyond the edge of the universe.

Regardless of where we are in the universe we still see galaxies moving away from us. But if galaxies are falling toward the center, how can they all be moving away? The answer is quite simple. It is vital to distinguish the difference between the direction of motion and the direction of acceleration. The direction of motion is toward the outer rim of the universe – up. The direction of acceleration due to expanding space is toward the center of the universe – down. It's like throwing a ball up in the air; the ball is moving up and accelerating down at the same time. We've assumed that because galaxies were moving away from us that they were also accelerating away from us; that assumption is false. Galaxies begin by *moving* in one direction and *accelerating* in the other.

Before we drop something from a high tower, the object is not moving relative to the Earth. Once the object is released it begins to pick up speed, moving in the same direction as its acceleration. This is the way we have become accustomed to thinking about free falling objects. So it is natural for us to think that as galaxies recede they are also accelerating away from us. But that is not the case; the situation is a little different with galaxies.

Suppose a galaxy is "dropped" from the top of the universe; it is already traveling at super high speed – near the speed of light. But from the time the galaxy is "released" it begins to accelerate toward the center of the universe. This is the same as decelerating from the direction it was moving. The galaxy would be in free fall, with its acceleration toward the center of the universe, the direction opposite the expansion. The galaxy would be traveling at high speed in one direction, but accelerating in the other. If this were true of all galaxies, then all galaxies are in free fall, decelerating from near light speed and accelerating (falling down) toward the center of the universe.

If this scenario were true, what would be our observations of galactic motion, and what would be different? We would still see younger galaxies moving away from us at a higher velocity than those closer to us. The rate of velocity increase would still be 70 km/sec/Mpc. All observations still hold true: the farther away the galaxy, the greater its velocity. No matter what our position in the universe, all scientific observations remain the same, only the interpretation has changed. But this interpretation changes everything! We have always assumed that galaxies and matter came from a single source – the singularity at the center of the universe. The assumption is that all matter has always existed and was blown into space by the big bang. That assumption is false!

In this chapter we have seen that all galaxies are in free fall. But now many questions have been introduced that need to be answered. Those answers will come in the following chapters. There is something amazing going on in the universe that we have never realized!

Chapter 25

The Expanding Universe

The Universe began with a Big Flash.

In the Beginning

We have seen Hubble's Law presented as evidence of the big bang theory. We have also seen that the big bang is subject to the laws of physics, except for the energy that sparked the big bang itself. If we reject the idea that all matter always existed and that there was no big bang, we are still left with the nagging question of where all of the mass in the universe came from. Let us now consider an alternative view of the creation of the universe and all the matter within.

Let's start with nothing: no matter, no light, no space, no time; absolute nothingness! Immediately this violates our current understanding of the laws of conservation of matter and energy. But if we start with nothing, then matter and energy must have come into existence at some point. Scientists readily admit that the nature of dark energy is not yet understood. The theory presented here is that dark energy is not subject to the laws of conservation. There is no fixed quantity; the amount of dark energy in the universe is growing at a steady rate, along with space, and has been doing so since the inception of the universe.

If we start with nothing, then something must have happened or else there would still be just nothing. What happened exactly we cannot know for sure, but based on what we see, we can make some logical inferences as to what happened. Suppose a hole was formed not in space and time, because such things didn't exist yet, but at the intersection of dimensions unknown to us. (Many theories postulate the existence of multiple dimensions beyond what we currently see, so this is not a stretch of the imagination. Multiple dimension theories abound, and the mathematics supports the possibilities.) This hole was the beginning of the physical dimensions which we call space and time, and in which our universe exists. Space and time are intimately connected, so much so we often call it spacetime.

This opening allowed pure energy to enter the new physical dimension. This new form of energy is what we now refer to as "dark energy," appropriately named because we can't see it. This is not speculation. Even though we can't see it, we know it exists because we witness its effects. We know there is an exotic force causing the expansion of the universe, a force different from those currently known to us. This is dark energy – the fundamental force that powers the expansion of the universe. Dark energy is exotic because it does not abide by the law of conservation of energy. Current speculation suggest that 70 percent of the universe is made up of this dark energy.[36]

Dark energy is a form of electromagnetic wave, therefore the speed of dark energy in empty space is the speed of light (2.9979×10^8 meters per second). The universe started to expand, and is still expanding in all directions at the speed of light. We can easily calculate the volume of space created in the first second by using the volume equation for a sphere.

$$v = \frac{4\pi r^3}{3} = \frac{4\pi (2.9979 \times 10^8)^3}{3} = 1.1286 \times 10^{26} \text{ cubic meters}$$

In the first second this volume of space became the entirety of the physical universe, that which is outside this volume is not part of the universe. Nothing physical exists outside of the universe, not even space and time, which are inseparable from the physics of this universe. Inside the universe we have space, time and energy. Outside of the universe there is nothing physical.

This of course means that the universe has a definite shape and size, and is growing in all directions. After one second of our scenario, our new universe contains only space, time, and dark energy. Within this first second there is no such thing as the force of gravity, the electromagnetic force, the strong nuclear force, or the weak nuclear force. Only dark energy – the fundamental force exists. It is the cause of expanding space, and thus the expansion of the universe. This expansion has been happening since the beginning of time, and will continue as long as time exists.[37]

Another Look at The Gravitational Constant

Before continuing with our description of the expanding universe let's take another look at the gravitational constant.

$$G = \frac{6.6742 \times 10^{-11} \ m^3}{kg * sec^2}$$

In the chapter on Space Consumption we saw that we could interpret the gravitational constant to be a measurement of a specific volume of space consumed by a kilogram of mass per second squared. We saw that the volume of space consumed every second by a kilogram of mass was 8.3871×10^{-10} cubic meters per kilogram, or .83871 cubic millimeters per kilogram.

Now let's look at the gravitational constant as a constant of space expansion. This would seem reasonable since we've already used the gravitational constant in our calculations of acceleration due to space expansion. It would appear that the gravitational constant provides important information concerning both the contraction of space and the expansion of space. Perhaps this is a case of an equal and opposite reaction. The driving energy is the fundamental force causing the expansion of the universe. The equal and opposite reaction may be the consumption of space by mass.

In the case of expanding space let's view the gravitational constant as the rate of increase in space without regard to any mass. This would make the gravitational constant a constant of proportionality with the following units of measurement:

$$G = \frac{6.6742 \times 10^{-11} \ m^3}{sec^2}$$

We saw in previous chapters that the rates of space contraction and expansion are based on kilograms of mass, so how can we now say that space expands "without regard to any mass?" At first Einstein believed the universe to be in a steady state, with mass always in existence. He did not think his field equation for relativity applied to a universe void of matter. However, De Sitter's solution to Einstein's equation showed that matter was not a necessary component. The universe could exist without matter.[38]

Let's consider the gravitational constant to be the rate of expansion of a volume of spherical space. Using the equation for the volume of a sphere we can calculate the rate of expansion in each direction.

$$v = 6.6742 \times 10^{-11} = \frac{4\pi r^3}{3}$$

Let's make some algebraic adjustments to calculate for "r":

$$r = \sqrt[3]{\frac{3 * 6.6742 \times 10^{-11}}{4\pi}}$$

$$r = 2.5163 \times 10^{-4} \text{ m/s}$$

This is the expansion rate per meter, in all directions from center, at the edge of the universe. This means that at the outer rim of the sphere of the universe, space is attempting to expand by 2.5163×10^{-4} meters for every meter of space. Multiply that by the speed of light and we get, 2.5163×10^{-4} * 299,792,458 = 75,436 meters per second. Each second this is the distance that space is attempting to expand *outside the bounds of the physical universe and beyond the speed of light.*

The Making of Matter

When the universe was first created there was no matter to begin with, and we're also saying there was no big bang. So where does matter come from? Einstein showed us that mass and energy are the same thing in different forms. The forms of matter and energy are interchangeable. It's like water and ice: same substance, different forms. To convert ice into water all we need to do is add energy – heat does nicely. To convert water into ice, all we need to do is remove some of the thermal energy from the water.

So it is with energy and matter. For now, let's think of matter as frozen energy. To convert energy into matter, some kind of energy must be applied. But when, where, and how does the conversion from energy to matter take place? A brief overview will follow, but the details of that conversion will be covered in a later chapter. For now, let's accept that this conversion takes place at the outer rim of the universe at nearly the speed of light.

Let's assume we are at the outer rim of the universe traveling at the speed of light. We see matter being created all around the outer rim of the sphere of the universe. We see energy being converted into electrons and protons. At the instant of creation the matter begins to fall toward the center of the universe at the rate of Hubble's constant. It would appear to us as if the matter were accelerating away from us. If that were the case then we would still see what we are seeing now. We would still see the galactic separation at the rate of the Hubble constant.

Although the scenario presented so far is different from the big bang theory, there are some similarities. In both cases the universe starts at one point and expands. In the big bang scenario the universe expands with an explosion and all existing matter is flung out into space, moving up toward the rim of the universe. This seems highly unlikely because the attraction of gravity is greatest when the matter is closest. The explosive blast of the big bang would have required an unfathomable amount of energy to disperse all existing matter throughout the entire universe.

In the alternate scenario presented here we have an expanding universe containing only energy within its realm of time and space. This dark energy is probably in the form of high energy electromagnetic waves, and would travel effortlessly throughout the expanding universe with no wasted energy.

One characteristic of this dark energy is that it appears to be spreading evenly throughout the universe without becoming diluted.[39] This dark energy expands as space itself expands. It appears as if the density of dark energy is the same everywhere throughout the universe. As the dark energy expands, space expands, and the universe also expands, but the energy density remains the same. It may be that dark energy is a characteristic of space itself.

One of the characteristics of space is that it expands at the rate of the gravitational constant. The universe itself is also expanding at the speed of light in all directions. Combine those two: the growth of the universe and the expansion of space within it, and we have another serious dilemma. We have space at the edge of the universe trying to expand outside the bounds of the universe itself, at a rate faster than the speed of light.

When space tries to expand faster than the universe itself something fascinating and wonderful happens. The wavelengths of the dark energy at the edge of the universe are compressed as they try to expand beyond the bounds of the physical universe, and beyond the speed of light. Just as the width of matter is compressed to near zero as it approaches the speed of light, the dark energy waves become so compressed that they are locked into a synchronous circular orbit. Thus energy is frozen, transformed into what we call matter. Once locked in place, matter is here to stay.

Electrons and protons are created at the outer perimeter of the universe. They are created from the energy in electromagnetic waves traveling at the speed of light. The electromagnetic waves are locked into a synchronous circular path. Thus, dark energy is converted into mass at very near the speed of light. This newly created mass is suddenly endowed with properties of mass and motion, such as momentum and inertia. This newly created mass is now subject to all the physical laws of matter. Physical mass cannot travel at or beyond the speed of light. So at the moment of creation its speed is just under the speed of light. The newly created mass obeys the laws of physics and begins to fall in the opposite direction from which it was traveling.

If we attempted to push an object of mass to higher and higher speeds strange things would happen as the object approached the speed of light. The amount of mass would begin to increase, its width would decrease, and its perception of time would be reduced. Forcing the object to move faster would take progressively more energy. At the speed of light the mass of the object would be infinite, its width would have shrunk to zero, and the perception of time would have come to a halt. This of course could not happen because it would take an infinite amount of energy to get any object of mass up to light speed. That's also why objects can't fall up toward the edge of the universe.

But what happens near the speed of light when electromagnetic waves become locked in a synchronous circular orbit and cannot travel at the speed of light which it was previously accustomed to? It is then that dark energy is transformed into matter. At very near the speed of light this newly created mass is traveling at incredibly high speeds, which gives it an incredible amount of inertia and momentum. The newly created matter would begin to fall as a result of space expansion. Thus the matter would immediately begin to slow down. The newly created mass would begin to accelerate toward the center of the universe. This would actually be a deceleration of speed. The rate of deceleration is 70 km per second per Mpc, which is exactly the rate we are seeing today.

The Conversion Rate of Energy to Matter

In the previous section we've seen that energy and the expansion of space have combined to produce matter. This only happens near the speed of light. The details of this process will come later. For now let's consider the rate of conversion of energy into matter.

We need to be able to calculate the new volume of space added to the outer rim of the sphere of the universe each second. In the chapter on Space Contractions, we saw Equation 8 – The Law of Volumes, used to calculate the flow of space. We can use that same equation here. It is shown again below:

$$v = 4\pi r^2 a$$

In this case:

    v   is the volume of space being created as the universe expands
    r   is the radius of interest
    a   is the rate of expansion of the universe – the speed of light

Let's calculate the volume of space by which the universe will increase each second at 14 billion years. At 14 billion years the radius of the universe will be about 1.3245×10^{26} meters. The velocity of expansion is the speed of light – 299,792,458 m/s. The volume of space created is:

$$v = 4\pi * (1.3245 \times 10^{26})^2 * 299{,}792{,}458 = 6.609 \times 10^{61} \text{ cubic meters}$$

The volume of space created in one second is 6.609×10^{61} cubic meters. At the same time, space is trying to expand at the rate of 75,436 meters per second beyond the rim of the universe. That expansion cannot happen. Let's do the same calculation to find the volume of space that could not be created:

$$v = 4\pi * (1.3245 \times 10^{26})^2 * 75,436 = 1.663 \times 10^{58} \text{ cubic meters}$$

1.663×10^{58} cubic meters is the amount of space that never got created. Instead the energy that would have occupied that would be space is converted into mass. The mass created is all hydrogen; nothing but protons and electrons. Space contains what is now called dark energy. This energy when converted into mass is equal to the density of hydrogen in space multiplied by the volume of space it occupies. This is shown in Equation 43:

Equation 43 – Mass Creation

$$m = dv$$

Where:
      m   is the amount of mass created
      d   is the density of hydrogen in space
      v   is the volume of space

The average density of space in our galaxy is about one lone hydrogen atom per cubic centimeter.[40] This would be the density of space where the hydrogen atoms did not coagulate into larger chunks. One hydrogen atom per cubic centimeter would be about 1.67×10^{-24} kg/m^3. For ease of calculation let's use a more conservative estimate of 1.0×10^{-24} kg/m^3. Using Equation 43, let's see how much mass was created in one second from dark energy during the expansion of the universe:

$$m = dv = 1 \times 10^{-24} * 1.663 \times 10^{58} = 1.663 \times 10^{34} \text{ kg}$$

For every 1.663×10^{58} cubic meters of space that cannot be created, there is 1.663×10^{34} kilograms of mass converted from the energy of that would be space. To simplify that let's say: for every 1×10^{24} cubic meters of space that cannot be created, there is 1 kg of mass converted from the energy of that would be space.

The Growth of the Universe

The table below shows the vital statistics of the growth of the universe over its first 14 billion years.

Table 30 – Growth of the Universe

Billion Years	Radius in Meters	Volume in Cubic Meters	Volume to Mass	New Mass in Kilograms	Total Mass in Kilograms
0	0.00E+00	0.00E+00	0.00E+00	0.00E+00	0.00E+00
1	9.46E+24	3.55E+75	2.68E+72	2.68E+48	2.68E+48
2	1.89E+25	2.84E+76	1.07E+73	1.07E+49	1.34E+49
3	2.84E+25	9.58E+76	2.41E+73	2.41E+49	3.75E+49
4	3.78E+25	2.27E+77	4.28E+73	4.28E+49	8.03E+49
5	4.73E+25	4.43E+77	6.69E+73	6.69E+49	1.47E+50
6	5.68E+25	7.66E+77	9.64E+73	9.64E+49	2.44E+50
7	6.62E+25	1.22E+78	1.31E+74	1.31E+50	3.75E+50
8	7.57E+25	1.82E+78	1.71E+74	1.71E+50	5.46E+50
9	8.51E+25	2.59E+78	2.17E+74	2.17E+50	7.63E+50
10	9.46E+25	3.55E+78	2.68E+74	2.68E+50	1.03E+51
11	1.04E+26	4.72E+78	3.24E+74	3.24E+50	1.35E+51
12	1.14E+26	6.13E+78	3.86E+74	3.86E+50	1.74E+51
13	1.23E+26	7.79E+78	4.53E+74	4.53E+50	2.19E+51
14	1.32E+26	9.73E+78	5.25E+74	5.25E+50	2.72E+51

The "Volume to Mass" column shows the volume of space that could not expand and must be converted into mass. The "New Mass in Kilograms" column shows the corresponding amount of mass created. The last column shows the accumulative total of the mass created. Every year more mass is created than in the previous year. According to the growth table, after 14 billion years the universe contains about 2.72×10^{51}

kilograms of mass. This matches very well with scientific observations and estimations.[41]

The Figure 48 shows the volume of the universe, and the amount of mass, being created over a time span of 30 billion years. The black line represents the volume of space in cubic meters, and the gray line represents the amount of mass in kilograms. The amount of mass created follows the volume of space being created at the rim of the universe.

Figure 48 – Volume and Mass of the Universe

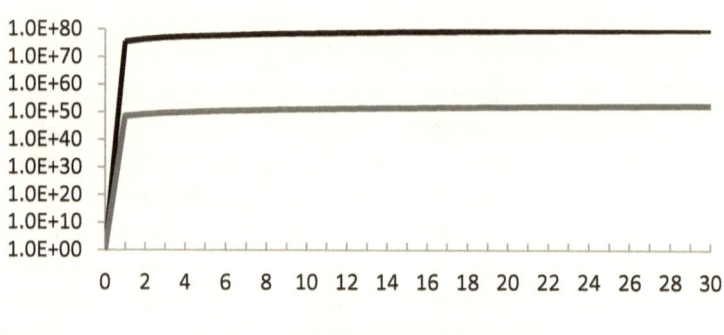

Time in Billions of Years

Some cosmologists theorize a period of rapid inflation in the early universe. This diagram shows a very rapid incline in volume and mass in the first billion years of growth. This is what we see when the size of the universe increases at the sure and steady speed of light.

Figure 49, shows the growth of mass in the universe over a trillion year period. On the very left of the graph we see the line representing the creation of mass going straight up. This line shows the amount of mass created in the first billion years, and it goes from zero to 2.68×10^{48} kilograms. The curve appears to be flattening, but not so. These are logarithmic graphs, so the growth is steadily inclining.

Figure 49 – Growth in Mass

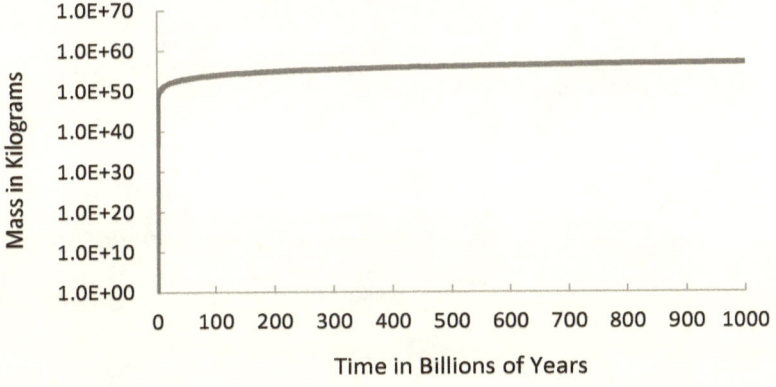

Perhaps Figure 50 gives a better illustration. In this graph the range scale of mass is smaller, where each level of the vertical axis is 10 times greater than the previous. The universe is growing very rapidly, at an exponential increase.

Figure 50 – Growth in Mass

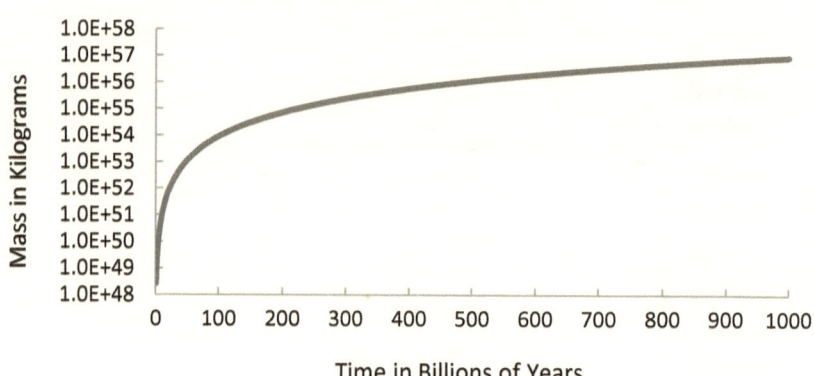

From these graphs we can see that the universe is growing exponentially, in both size and mass. As the universe grows at the speed of light in all directions, new matter is constantly being created at the top of the universe. Once created the mass begins to fall.

Figure 51 shows the density of mass in the universe over the first thirty billion years. After the first few billion years the density appears to level off. After 30 billion years the density is about 2.64×10^{-28} kilograms/cubic meter. The density is actually decreasing slowly, asymptotically approaching 2.5163×10^{-28} kilograms/cubic meter. The density of the universe is closely related to the gravitational constant. It would appear that the universe, still in its youth, will be in a steady state of growth forever.

Figure 51 – Mass Density of the Universe

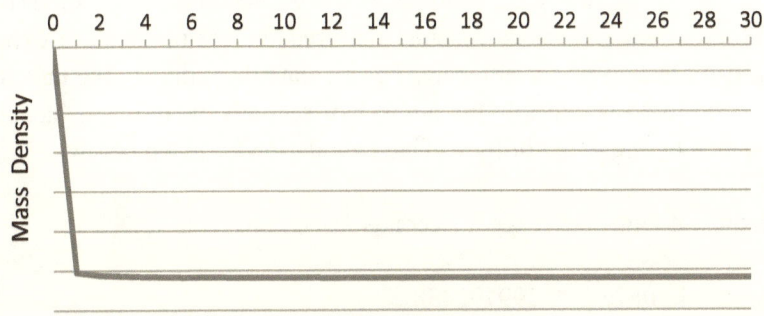

It would also appear that there is no end to the growth of the universe, or the creation of matter. If all continues as it has from the beginning there will be no end to the universe. The universe has no expiration date; it is here to stay.

Expansion Rate

In this chapter we have emphatically stated that the growth rate of the universe in all directions is the speed of light. As of yet no proof has been offered. The proof comes now. In the chapter entitled 'Hubble's Law' we saw that scientists believe that Hubble's Constant is a value that shows the rate at which galaxies are separating from each other. Through scientific observation Hubble's Constant has been measured to be approximately 70 Kilometers per million parsecs of space. We used Hubble's Constant to calculate the age of the universe. That calculation is repeated below.

Distance Between:	3.08609E+22	meters
Expansion Rate:	70,000	meters/s
	4.4087E+17	seconds

The age of the universe is about $4.4087e^{17}$ seconds. That converts to 13.97 billion years. That is the time required for the universe to expand from a Big Flash to its current state. To calculate the radius of the universe, all we need to do is multiply the number of seconds the universe has been in existence by the speed of light. That calculation is shown below.

Radius of Universe = age in seconds * speed of light.

$$R_u \; = \; 4.4087e^{17} \; * \; 299792458$$

$$R_u \; = \; 1.321695e^{26} \text{ meters}$$

Now, just one more calculation. Lets divide the radius of the universe by the time squared to get the acceleration of objects in the universe.

$$a \; = \; \frac{R_u}{T^2} \; = \; \frac{1.321695e^{26}}{(4.4087e^{17})^2} \; = \; 6.8000e^{-10} \quad \frac{m}{s^2}$$

The acceleration of objects in space is exactly the same as calculated from Hubble's Constant in the previous chapter. There is no doubt; the universe is expanding as described in this chapter, and the galaxies are in freefall as described in the previous chapter.

Chapter 26

The Making of Galaxies

Galaxies fall like rain!

The Rain of Galaxies

We previously suggested that galaxies are falling from the top of the universe toward the center. The following analogy is used to help us envision how that happens.

We've stated that matter is simply frozen energy. Let's compare the conversion of energy into matter, with vapor and water. The Earth is covered with an abundance of water. Apply energy to the water in the form of sunlight and a conversion takes place. Water is transformed into vapor. Water vapor has different characteristics than water even though it's made of the same substance. Water vapor is light and can readily rise to the upper regions of the atmosphere. The vapor drifts up without being heard or seen. In the upper regions of the atmosphere this vapor has a tendency to gather itself together to form clouds. Eventually, when conditions are just right, the water vapor changes back to its original form. As water, it falls back to Earth.

This is similar to what happens with the conversion of energy into mass. Having no mass, dark energy travels uninhibited toward the upper regions of the universe. At the edge of the universe dark energy is converted into mass. Mass has different characteristics than energy, even though they're composed of the same stuff. Immediately following its transformation into mass, in the form of hydrogen, it begins to fall toward the center of the universe. The hydrogen clumps together to form stars within the galaxies. The galaxies fall like rain! Just like raindrops and snowflakes, all the galaxies have different forms, shapes and sizes; no two are exactly alike.

Keeping this analogy in mind, let's look at the mechanism in which galaxies are made.

Galaxies in the Making

The universe is expanding at the speed of light. Matter in the form of hydrogen gas is frozen into being at the outer perimeter of its sphere. The top being the outer perimeter of the universes sphere. As we have previously calculated, the additional space created along the outer perimeter of the universe, in one second, is 6.609×10^{61} cubic meters, assuming the universe is about 14 billion years old. Due to the compression and transformation of dark energy, that space is filled with 1.663×10^{34} kilograms of hydrogen gas. This is an ongoing process. The density of space in the outer perimeter is 2.5163×10^{-28} kg/m^3. See the calculation below.

Mass created in 1 second	1.663×10^{34}	kilograms
Volume created in 1 second	6.609×10^{61}	cubic meters
Density of space	2.5163×10^{-28}	kg / m^3

Once created from energy at the top of the universe, matter begins to fall toward its center. In the process of falling, clouds of hydrogen gas are formed. The gravitational attraction of the newly created hydrogen atoms give the matter a tendency to clump together. At first, these clumps would be no larger than specks of dust or sand. Those specks would then coalesce into larger chunks the size of raindrops and hail. The sizes of the accumulation of matter grow increasingly larger. Eventually there are billions of massive bodies of hydrogen circling larger structures of mass in the center. This huge conglomeration of mass takes on the shape and form of what we call a galaxy.

Any slight imbalance like the gravitational pull of a neighboring galaxy and the effect of space expansion, would cause the newly forming galaxy to rotate on its journey down from the top of the universe. Thus all galaxies rotate, as well as the newly created bodies within them. Like water down the drain, everything rotates as it falls.

Within the galaxy these huge bodies of hydrogen continue to coalesce. When large enough, they ignite, and stars are born. These early stage galaxies are referred to as star nurseries, and within them stars are born by the millions. Within these stars elements are created in the furnaces of nuclear fusion. First come neutrons, which are an electron and proton squashed together. These neutrons are the key ingredient necessary for holding the nucleus of atoms together. Then hydrogen atoms combine to form helium atoms. The combining goes on to form lithium, beryllium, boron, carbon, nitrogen, oxygen, etc.

The largest stars in the center of the galaxy burn out the fastest, ending their lives with a bang – a supernova. The elements that had been created with these stars are then scattered throughout the galaxy. All the elements in our bodies, with the exception of hydrogen, were created within the internal inferno of a star. Some of the elements, heavier than iron, are created in the supernova explosion at the end of the star's life. Much of this debris from the supernova is captured by other stars within the galaxy.

But remember: space is in a constant state of expansion, and the effects of this expansion are ever present. The debris captured by the stars may have a highly elliptical orbit at first. But as time goes by the orbits become ever more circular, and the debris begins to coagulate. Eventually planets, moons, asteroids and comets are formed throughout the galaxy.

Matter is born at the top of the universe and immediately begins to fall toward the center. The rate of the fall has been measured to be about 70 km/sec/Mpc. This is the Hubble constant and this is what we now see. We still see the separation of the galaxies. They are still receding, but they are also decelerating from the speed of light, accelerating away from the outer rim of the universe – toward the center. This means that we cannot use Hubble's law as proof of the big bang, or as a means to determine the exact age of the universe.

Hitting Bottom

As previously mentioned, galaxies are receding from us, but also accelerating toward us. Eventually, from our perspective, galaxies will stop receding, come to a halt, then begin to accelerate toward us. One may question what happens when the falling galaxies hit bottom, when galaxies coming from all directions finally reach the center of the universe? As surprising as it may sound, galaxies can pass through each other without a single collision of the stars within. So widely distributed are the stars in a galaxy that the chances of any two meeting head on would be highly unlikely. The galaxies could pass center and continue on their course. However, once past center, the acceleration of a galaxy would reverse direction. Remember, the acceleration is in the opposite direction of space flow. Once a galaxy crossed center its acceleration would reverse and it would then begin to slow down. Of course it would go a long way before actually coming to a stop and reversing direction. If that happened over and over again, all galaxies would eventually end up in the center of the universe in the biggest black hole to ever exist. I strongly suspect that that will not happen.

Consider the planets: all are attracted toward the center of the solar system yet they never fall into the Sun. Consider the stars: all are attracted toward the black hole at the center of the galaxy, yet they are never pulled into it. The planets orbit the Sun, and the stars orbit the center of the galaxy. So it is with galaxies. As attracted as they are toward the center of the universe, they will never arrive. They will all end up in orbit around the center of the universe. The galaxies behave just like planets within a solar system and stars within a galaxy, revolving in orbit around the center of their attraction. The galaxies establish orbits for the same reason that planets orbit their Suns, and stars orbit the center of their galaxies. The same laws apply to all.

One significant difference between the orbits of planets and those of galaxies is that the planets revolve around the Sun on about the same plane of axis. This is also true of the stars within a galaxy. Most galaxies are shaped like large disks, so the stars also rotate around the center of the galaxy on about the same plane. With galaxies it is different. Because the galaxies are all approaching center from different directions, they will orbit in different directions, with highly elliptical orbits at first. Those orbits will eventually become circular just like the planets.

Cosmologists do not know how long galaxies live before exhausting all their fuel. This would be hard to determine because we don't know how much matter is still floating around the galaxy in the form of dust not yet coagulated into stars. And we don't know how much matter is still floating around the outer rim of the galaxy, and in the galaxy's halo, because we can't see it. But there is still enough conglomeration going on in the Milky Way to give birth to 1 to 3 new stars every year.[42]

Will falling galaxies collide in the center of the universe, or will they establish orbits? We will know eventually – in about 3-5 billion years. The Andromeda and Milky Way galaxies appear to be rushing toward each other.

Figure 52 – Andromeda

Credit: NASA/JPL–Caltech

http://www.galex.caltech.edu/media/glx2008-01f_img01.html

There is not enough information now to know whether they are on a collision course or if they will end up rotating around each other. We'll just have to wait and see. A collision would cause more conglomeration, which would give birth to many new stars. Collisions and close encounters revive aging galaxies.

If we consider the universe to be 13.7 billion years old,[43] and the age of the Milky Way to be 13.6 billion years old[44], we would have to consider the Milky Way among the first galaxies ever to exist. Both the Milky Way and the Andromeda galaxies show signs of having already merged with other galaxies. If so, we might consider ourselves to be located very near the center of the universe. Will the Milky Way and Andromeda begin the galactic rotation around the center of the universe? Or will they eventually merge to create one super large spiral galaxy? Only time will tell.

The Universe According to Hoyle

English astronomer Sir Fred Hoyle (1915-2001) was a critic of the Big Bang Theory, and is believed to have coined the term "Big Bang." Hoyle saw the theory as an irrational process that could not be described in scientific terms.[45] There were, and still are, many issues with the Big Bang that have caused concern among physicists. Some of these questions still remain unanswered. We explained them away by acknowledging that we don't yet understand the physics of a singularity. Much is accepted on faith because there are no simple answers to the really tough questions.

Before the Big Bang Theory became widely accepted by cosmologists, Hoyle and fellow scientists Hermann Bondi and Thomas Gold proposed a "steady state" model of the universe.[46] This model accepted the expansion of the universe and the recession of galaxies, but not the Big Bang itself. Instead of all matter being created at once, Hoyle and his colleagues suggested that matter is created continuously, a little at a time, in the vast spaces between receding galaxies. So as galaxies receded from each other, new ones came into existence in between. Thus everything is kept in a steady state.

Hoyle had no convincing argument for the spontaneous creation of matter between galaxies, so the theory was not widely accepted. When cosmic microwave background radiation was discovered in 1965 by Arno Penzias and Robert Wilson of Bell Laboratories, proponents of the Big Bang claimed that as evidence of the Big Bang. According to Penzias and Wilson, this background radiation was "left over" from the Big Bang. Hoyle could proffer no explanation for the background radiation, so the Big Bang Theory obtained credibility and is now the dominant theory of how the universe began. But, is cosmic microwave background radiation really evidence of a Big Bang?

Information recently obtained by scientists using NASA's Wilkinson Microwave Anisotropy Probe (WMAP) shows that the cosmic microwave background radiation comes to us after traveling over 13 billion years. What we see in Figure 53 is the edge of the universe 13 billion years ago. In it we see patterns of cosmic radiation scattered throughout the whole universe. These patterns correspond with the birth of galaxies. "Within this light are infinitesimal patterns that mark the seeds of what later grew into clusters of galaxies and the vast structure we see all around us."[47]

Figure 53 – Cosmic Background Radiation

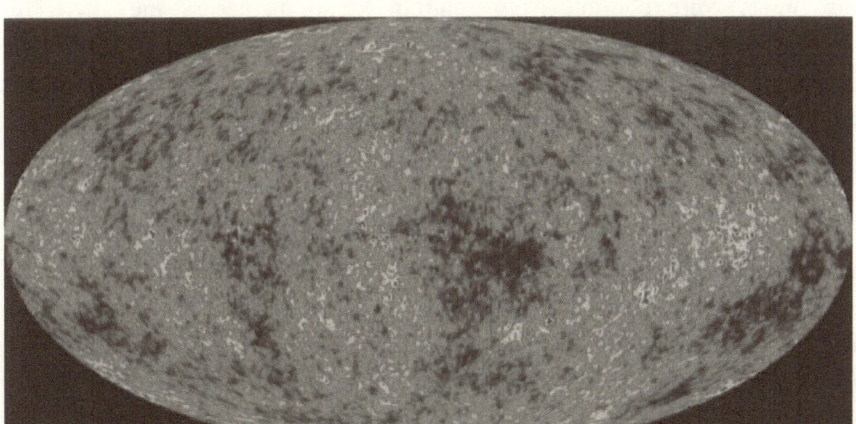

Credit: NASA/WMAP Science Team

http://www.nasa.gov/centers/goddard/images/content/96115main_Full_m.jpg

What we really see in Figure 53 is the pattern for the birth of galaxies. Is the cosmic microwave background radiation really the afterglow from the Big Bang, or is this cosmic radiation generated as the result of new galaxies in the making? The cosmic microwave background radiation may be coming from the conversion of hydrogen to helium. Much of what is given as proof of the Big Bang theory could better be seen as evidence of matter being created at the outer rim of an expanding universe.

Gamma-ray Bursts

In the late 1960s, nuclear detection satellites observed something unusual from outer space. Short but brilliant flashes of light and high energy gamma-rays. This phenomena came to be known as gamma-ray bursts. The questions were: what are they, what causes them, and where do they come from. This phenomena peaked the curiosity of cosmologists. But information was hard to come by. The bursts were of short duration, measured in seconds, so capturing sufficient data to work with took decades.

At first astronomers thought these gamma-ray bursts were occurring within our galaxy. This was a logical assumption because of the enormous amount of energy produced by the bursts. However, observations showed the bursts were coming from random locations across the sky. This suggested that the bursts originated outside the galaxy. If they were occurring within the Milky Way they would happen in the galactic plane. This was not the case. The problem of origin sparked even greater interest because of the enormous amounts of energy they produced.

After several decades of observations the story is just beginning to come into focus. The gamma-ray bursts originate from supernova, the kind of supernova from which black holes are created. These supernova and resulting gamma-ray bursts are happening in deep space. They occur from 0.5 billion to about 12.3 billion years away from us; with most about 9.3 billion years away.[48] This puts them at the edge of the visible universe. The gamma-ray bursts can be seen over such great distances because they are the brightest objects in the universe. The farthest gamma-ray burst yet observed is only about 1.7 billion light years from the actual edge of the universe.

Visualize what is happening. All around us in every direction, galaxies are being formed in the outer regions of the universe. These young galaxies are giving birth to multiple millions of stars. Supernova occur most frequently in these young galaxies having very little metal content. The supernova at the centers of these galaxies are creating the black holes to anchor the rotation. Meanwhile, the high density elements created from within the exploding stars are scattered throughout the galaxy. These heavier elements will become the future planets and moons of the galaxies.

Chapter 27

The Nature of Matter

Wave or particle?

Electromagnetic Wave Review

In the preceding chapters we postulated that matter is created from energy at the top of the universe. We only offered some analogies to help illustrate how that transformation takes place. The explanation of how energy is converted into matter comes in the following chapter. Before that explanation comes, this chapter will highlight some of the interesting discoveries of the nature of matter.

Before discussing the nature of matter, let's review what we know about electromagnetic waves. Electromagnetic waves are pure energy. They have no mass and travel at a fixed speed in empty space. That fixed speed is the speed of light: 299,792,458 meters per second. That's the speed limit regardless of which direction it's going. That's the speed limit regardless of the speed and direction the source of the energy is traveling. That's the speed limit no matter what kind of electromagnetic wave it is: light wave, radio wave, microwave, etc.

All electromagnetic waves travel at the same speed. The difference in electromagnetic waves is in their frequency and wavelength. Frequencies and wavelengths are inversely related. Frequency times wavelength is equal to the speed of light. The range of frequencies or wavelength is referred to as the electromagnetic spectrum. Electromagnetic wavelengths are also related to the energy they carry. Short wavelengths have a higher frequency and carry more energy than longer wavelengths.

Electromagnetic waves are classified by their frequency. The following is a list of electromagnetic waves according to their relative frequency. They are listed in order of lowest frequency (longest wavelength) to highest frequency (shortest wavelength): radio waves, microwaves, infrared light, visible light, ultraviolet light, X-rays, and gamma rays. Somewhere in this list, or beyond this list is the elusive dark energy that pervades the whole universe. Visible light waves are just a small part of the whole spectrum.

Electromagnetic waves are self propagating waves. They travel through empty space one step at a time. Since there are only two "feet", electric waves and magnetic waves, we can call it the Texas Two Step. The motion of the electric wave will generate a magnetic wave. Then the motion of the magnetic wave will generate an electric wave. It's just two steps, but the cycle constantly repeats itself over and over again. The frequency of this cycle is the frequency of the electromagnetic wave.

Young's Double Slit Experiment

In the early 1800's English physicist Thomas Young (1773-1829) conducted one of the most fascinating experiments ever performed. Now referred to as The Double Slit Experiment, Young's objective was to prove conclusively whether light traveled as a wave or a stream of particles. Newton had suggested that light was transmitted as tiny particles. Young believed light to be a wave.

Young projected a light through two slits in a wall in order to observe the display of light patterns on the wall on the opposite side of the slits. American physicist Richard Feynman describes the experiment as firing bullets through two holes in an armor plated wall.[49] Any bullets to make it through the wall must have traveled through the two holes. The resulting pattern of bullet holes on the second wall would clearly show the path the bullets had traversed. If light travels as particles, the expectation was that the pattern of bullet holes on the second wall would be in line with the holes in the first wall as well as the rifle that had fired the bullets. Figure 54 depicts the expected pattern when light waves, behaving as physical particles, are fired at two slits in a wall.

Figure 54 – Particle Pattern

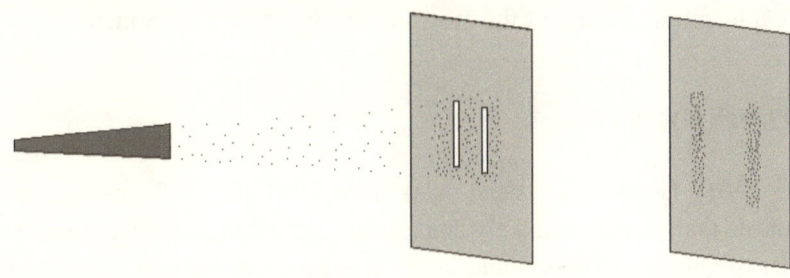

If light travels as a wave, the pattern on the opposite wall would be very different. In this case the analogous example is of a calm pool of water with a rock tossed in the middle. Waves are generated outward from center in concentric circles. The retaining wall has slits instead of holes that permit the waves to pass through simultaneously. The waves emerging from the slits on the other side will spread out and interfere with each other. If light travels as a wave then the resulting pattern on the opposite wall would be an interference pattern as shown in Figure 55.

Figure 55 – Interference Pattern

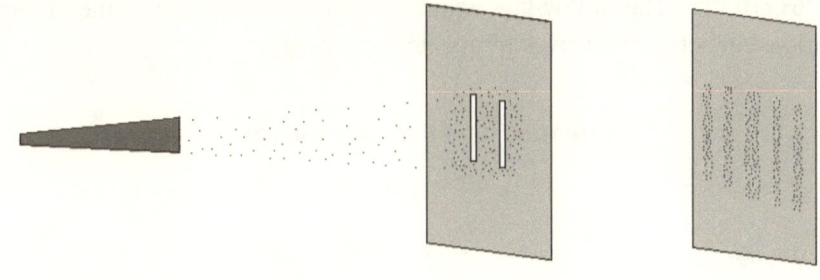

Young's experiment showed the interference pattern. Light travels as a wave. The experiment appeared to be conclusive and satisfying to the English physicist. But our understanding is still incomplete, the story is not over yet. In the first quarter of the 20th century there was a rapid growth in our understanding of matter and electromagnetic waves. Following is a brief explanation of some of the most interesting related experiments.

The Planck Hypothesis

In 1901, German physicist Max Planck (1858-1947) presented a hypothesis that would ultimately change the face of physics. Plank was the founder of Quantum Theory, and because of his work we now have a field of science called Quantum Physics. While conducting experiments on black body radiation, Planck realized that electromagnetic radiation comes in little bundles of specific quantity, called quanta, now known as photons. At the time Planck presented his thesis, the very concept of quantum physics was difficult to understand. As a new science in its infancy quantum physics took time to gain in popularity, but mounting evidence proved his theory, and in 1918 Planck was awarded the Nobel prize in physics for the discovery of energy quanta.

Planck also discovered a direct relationship between the frequency of electromagnetic radiation and the energy contained within. The higher the frequency the greater the energy. The lower the frequency the less energy. The energy of an electromagnetic wave divided by its frequency is a fixed number called Planck's Constant. According to the National Institute of Standards and Technology (NIST), the value of Planck's constant is 6.6261×10^{-34}.[50] The following equation from Planck shows the direct relationship between a waves frequency and energy:

Equation 44 – Planck's Energy

$$E = hf$$

Where:

E   is energy

h   is Planck's constant

f   is wave frequency (defined as c ÷ wavelength (λ_c))

Planck's constant can also be thought of as the mass of an elementary particle multiplied by its Compton wavelength (represented by the Greek character lambda – λ_c) times the speed of light. This relationship is shown in the following expression: $h = m\lambda_c c$. The calculation for Planck's constant is shown below using the mass of an electron:

$$h = m\lambda_c c = 9.1094 \times 10^{-31} * 2.4263 \times 10^{-12} * c = 6.6261 \times 10^{-34}$$

We can also use Equation 44 to calculate the energy in the elementary particles of matter. First we'll substitute "$m\lambda_c c$" for h; then we'll substitute "$c \div \lambda_c$" for frequency "f". We can follow the transformation to the new equation below:

$$E = hf = m\lambda_c cf = m\lambda_c c * \frac{c}{\lambda_c} = mc^2$$

Of course the equation $E = mc^2$ looks very familiar. This shows that Planck's energy equation is applicable to both electromagnetic waves and matter. This also indicates that matter and energy may be more closely related than we realize. We will be using Equation 44 again shortly to make a very important point concerning matter and energy.

Einstein's Photoelectric Effect

In 1905 (the same year he published his paper on the Special Theory of Relativity) Einstein published another paper describing a peculiar phenomenon now known as the photoelectric effect. The significance of his paper was not realized until much later as the understanding of quantum physics became more prevalent. It wasn't until 1921, 16 years after its publication, that Einstein was awarded the Nobel prize in physics for the discovery of the photoelectric effect.

Einstein observed that electrons could be dislodged from some metals by shining light on them. The characteristics of these electron emissions could not be explained if the incoming light was a wave. Only if the light striking the metal came in little tiny packets of energy could Einstein explain the photoelectric effect. These tiny packets of light are what we now call photons. Einstein's observations did not mesh with Young's double slit experiment that showed light to be an electromagnetic wave. But, the observations did coincide perfectly with Planck's hypothesis of energy quanta. Einstein was not a proponent of quantum physics, but it was this discovery that established it securely.

Once again the nature of light was brought into question. Is light a wave, a particle, or both? The story is still not finished.

The Atomic Model

In 1909 English physicist Ernest Rutherford (1871-1937) presented a model of the atom which appeared very much like a miniature solar system. His model had a dense nucleus surrounded by a cloud of orbiting electrons. This model was not readily accepted by the scientific community because of its instability. The prevailing understanding of physics at that time was that the electron continually emits electromagnetic radiation. The thought regarding Rutherford's atomic model was that the electron would quickly radiate its energy, then collapse into the dense nucleus. An atom's life would be short lived.

The Bohr Radius

In 1913, Danish physicist Niels Bohr (1885-1962) presented a model of the atom that solidified the quantum theory. It was this model that won him a Nobel prize in 1922.

Bohr studied the light emitted from hydrogen gas when energy was applied. The light emitted was not in the continuous range of wavelengths of visible light, but rather was emitted in specific colors. Other gases showed the same effect, but with different colors. This peculiar phenomenon was known for years before Bohr began his investigation, but until then remain unexplained. Bohr translated his discoveries into a rectified version of Rutherford's atomic model that proved to be stable.

Bohr postulated that the orbits of the electrons were allowed at only very specific energy levels. Within a stable orbit, the electron would radiate no electromagnetic energy. Thus the electron could maintain its energy level without spiraling into the nucleus of the atom. Light waves are emitted only when an electron transitions into an orbit of a different energy level. A different wavelength is emitted during a transition to each of the possible orbits. The wavelength emitted is related to the difference in energy levels of the two orbits involved in the transition.

The specific electron orbits are associated with specific energy levels. The development of quantum theory was now underway. In order for an electron to transition to another orbit it had to make a "quantum leap." An electron transitions to another orbit whenever a quantum amount of energy is absorbed or emitted from an electron in the atom. But the transitions are always to a stable resonant orbit.

Planets can revolve around the Sun at any distance. But this is not so in the quantum world of atoms. The stable orbits of electrons are standing waves that only occur at resonant frequencies. An electron's frequency is equal to 2π times the radius of the electron's orbit. At the lowest energy level an electron will make one revolution at the lowest resonate frequency. Each succeeding higher orbit will be an integer multiple of the first (lowest) energy level orbit.

The radius of an electron's orbit at the lowest energy level is referred to as the Bohr radius. It is 5.2918×10^{-11} meters. The radius at the next highest energy level would be $2 * 5.2918 \times 10^{-11}$ meters. The next would be $3 * 5.2918 \times 10^{-11}$ meters, and so on. The energy levels and orbits of electrons are quantized by discreet amounts.

Louis de Broglie's Amazing Hypothesis

In 1924, just two years after Bohr received his Nobel prize, French physicist Louis de Broglie (1892-1987) presented in his doctoral thesis the theory of electron waves. De Broglie must have been considering the dual nature of light, and thought that perhaps matter also had such a nature. He hypothesized that matter is actually made up of waves or had wave like characteristics. This would help explain why electrons can have only certain orbits. He proposed that the wavelength of an electron is equal to the Planck constant divided by its momentum. Equation 45 illustrates this relationship:

Equation 45 – Wavelength of Particles

$$\lambda = \frac{h}{mv}$$

Where:

  λ is the wavelength of a particle
  h   is Planck's constant (6.6261×10^{-34})
  m   is the mass of a particle
  v   is the velocity of a particle (c * fine structure constant)

Let's use this equation to calculate the wavelength of an electron:

$$\lambda = \frac{h}{mv} = \frac{6.6261 \times 10^{-34}}{9.1094 \times 10^{-31} * c * 7.2974 \times 10^{-3}} = 3.3249 \times 10^{-10}$$

The wavelength of an electron is 3.3249×10^{-10}. Divide it by 2π to get the radius of an electron - 5.2918×10^{-11} meters. This turns out to be the Bohr radius.

De Broglie was awarded the Nobel prize in physics in 1929 for the discovery of the wave nature of electrons. This added support to Bohr's hypothesis. But, even more importantly, De Broglie's brilliant insight gave us the understanding that elementary particles also have wave-like characteristics.

The Davisson-Germer Experiment

In 1927, three years after the De Broglie hypothesis, two Bell Laboratories scientists: Clinton Davisson and Lester Germer tested his hypothesis. Would electrons have the same diffraction patterns exhibited by light? This is a phenomenon believed to be displayed only by waves. Who would have expected the surprising results? Yes, the diffraction properties of reflected electrons is the same as light.

The experiment conducted by Davisson and Germer involved only electrons. Since then, experiments have confirmed that other elementary particles also exhibit wave-like properties. De Broglie's hypothesis has been confirmed. The fundamental particles, the building blocks of all atoms: electrons, protons, and neutrons, all have wave-like properties.

Sophistication was added to these experiments as technology improved. The next step was to perform the experiment with only a single photon at a time. Would the resulting pattern be a group as would result from the bullet analogy, or would single photons fired one at a time show an interference pattern? The results showed the same interference pattern. This meant that a single photon traveled through both slits at the same time. The experiment was then performed with a single electron. The results were the same. It would appear that a single electron passes through both slits at the same time. American Physicist Richard Feynman says that this is absolutely impossible to explain with classical physics. "In reality, it contains the *only* mystery. We cannot explain the mystery in the sense of 'explaining' how it works."[51]

So we've seen electromagnetic waves with their dual nature displaying wave-like and particle-like properties. Now we see the fundamental particles of matter also displaying a dual nature. The building blocks of atoms have wave-like properties. These experiments, conducted by extraordinary individuals, resulted in extraordinary discoveries. These discoveries are clues to the creation of matter.

Chapter 28

The Creation of Matter

"Let there be light"   Genesis 1:3

Energy of Matter and Waves

Now that we've gone through a review of several of the most interesting discoveries in science we can begin putting all this information together. We can think of electromagnetic waves as being analogous to waves of water. The water in a pool is made from billions and billions of tiny water molecules. A stone thrown into the pool will cause the oscillation of the water molecules and outward flowing waves are created in concentric circles. Electromagnetic waves are similar, only they flow outward in three dimensions. These waves can also be thought of as being made from many extremely tiny particles of light called photons.

First let's start by using Equation 44 – Planck's Energy equation to calculate the energy of an electromagnetic wave of a specific wavelength. Remember that frequency is defined as the speed of light divided by its wavelength. In this case we'll divide the speed of light by Compton's wavelength of the electron to get the frequency.

$$E = hf = 6.6261 \times 10^{-34} * c \div 2.4263 \times 10^{-12}$$
$$E = 8.1871 \times 10^{-14} \text{ joules}$$

The energy of a photon of an electromagnetic wave having Compton's wavelength is equal to 8.1871×10^{-14} joules. Now, let's calculate the energy of an electron. The energy of an electron is its mass times the speed of light squared. Let's do the math:

$$E = mc^2 = 9.1094 \times 10^{-31} * c^2$$
$$E = 8.1871 \times 10^{-14} \text{ joules}$$

Here we see that the energy of a photon with Compton's wavelength has exactly the same energy as an electron. One is a particle of an electromagnetic wave the other is a particle of mass. Their energy levels are the same even though their wavelengths are different. Could they be the same thing in different forms?

Let's take a look at some practical examples of energy exchanges before getting into the mathematical analysis. Einstein's discovery of the photoelectric effect lead to the invention of solar cells which people use today to provide power to their homes. The solar cells are arrayed into panels that are mounted on rooftops of homes and configured to face the Sun. The electromagnetic waves from the Sun are converted into electrons in the solar cells. This process generates electricity which provides power for their homes.

The same thing happens in the Earth. As the Sun shines its light on the Earth, the Earth absorbs the electromagnetic radiation and stores the energy in its atoms. After a day basking in the sunshine the Earth is charged with energy. On any given summer night, overlooking the mountains of Colorado, for hours into the night, the accumulated energy can be seen discharging in the form of electrons in a dazzling display of lightning throughout the mountains.

In both cases, electromagnetic radiation is being converted into electrons. The electrons and electromagnetic waves seem to be very closely related. And this conversion works both ways. Consider electricity traveling through a copper wire. Copper is used for wiring because of its abundance of free electrons. Electricity travels through a wire at the speed of light, but electrons can't do that. What's going through the wire are electromagnetic waves. But where do they come from? They must come from the electrons within the wire.

Watching the flickering dance of flames in a fireplace on a cold winter's evening can be a very warming and relaxing experience. The combustion of the wood creates radiation of two kinds. Heat radiates out from the fireplace as well as light. The light is generated as electrons are constantly shifting up and down to different levels of Bohr radii. This is how scientists can tell what elements are in the stars. The spectrum of colors are matched with the colors radiated from various elements. Each element has a specific color signature.

These common occurrences beautifully illustrate the dual nature of matter. These examples to not create or destroy energy or matter. They simply illustrate the common conversions of matter into energy and energy into matter. The study of these interactions between electromagnetic waves and matter is known as quantum electrodynamics. We are about to investigate a curious little number that comes up over and over again in the computational analysis of quantum electrodynamics.

The Fine-Structure Constant

The radius of an electron's orbit around the nucleus of an atom at the lowest energy level is the Bohr radius (5.2918×10^{-11} meters). The classical radius of an electron is 2.8179×10^{-15} meters. When doing the comparison of energy levels in the previous section, we compared the energy of an electron with that of a photon with Compton's wavelength. Dividing Compton's wavelength by 2π gives us the Compton radius which is equal to 3.8616×10^{-13} meters. These are well known numbers among physicists and are often used in mathematical computations. But what does it all mean? Why the different size radii and how are they related?

In 1916 German physicist Arnold Sommerfeld (1886-1951), discovered the relationship between the velocity of an electron in orbit and the speed of light. This is a constant value now called the fine-structure constant. The mathematical symbol for the fine-structure constant is the Greek letter alpha "α", and its NIST value is $7.2973525376 \times 10^{-3}$. The inverse of the fine-structure constant is also often used in mathematical calculations and its NIST value is 137.035999679.[52]

The radii discussed above, as well as their equivalent wavelengths are also related by the fine-structure constant. Let's take a look at the difference between these radii. Table 31 illustrates the relationships. The Bohr radius multiplied by the fine-structure constant is equal to Compton's radius. Compton's radius multiplied by the fine-structure constant is equal to the classical electron radius. Of particular importance is the relationship between the Bohr radius and the classical electron radius. The classical electron radius divided by the fine-structure constant squared is the Bohr radius.

Table 31 – Radii Relationships

Radii	Relationship	Values
Bohr radius	$5.2918 \times 10^{-11} * \alpha =$	3.8616×10^{-13}
Compton radius	$3.8616 \times 10^{-13} * \alpha =$	2.8179×10^{-15}
Classical electron radius	$2.8179 \times 10^{-15} \div \alpha^2 =$	5.2918×10^{-11}

The fine-structure constant is a number that was discovered because of the relationships between the speed of light and the speed of an electron, and the difference in the radii previously mentioned. This is a number derived from relationships. In fact, it is defined by many relationships. It is sometimes referred to as a coupling constant because it is used in calculations regarding the exchange of photons in quantum electrodynamics. But no one knows where this number comes from. No one knows its origin. It has remained a great mystery of science since its discovery almost 100 years ago.

American physicist Richard P. Feynman (1918-1988) said: "... all good theoretical physicists put this number up on their wall and worry about it."[53] Here's another quote from Feynman: "It's one of the *greatest* damn mysteries of physics: a *magic number* that comes to us with no understanding by man. You might say the 'hand of God' wrote that number, and we don't know how He pushed His pencil."[54] We are about to see the writing on the wall!

Two Different Electrons

The difference between an electron with the Bohr radius and an electron with the classical electron radius is the fine-structure constant squared. That relationship is shown in the following equation:

$$\text{Bohr radius} = \frac{\text{classical electron radius}}{\alpha^2}$$

Which can also be written as:

Bohr radius * α^2 = classical electron radius

But, are there really two different electrons? Let's illustrate that by using the acceleration equation shown below. The derivation of this equation will be shown in a later chapter. In this case we've dispensed with the distance squared in assuming interaction at a distance of one meter. If we apply this equation to an electron, then this equation tells us that the acceleration of an electron due to another charged particle one meter away is equal to the radius of the electron times the speed of light squared.

Equation 46 – Acceleration by Charge

$$a_x = \frac{r_x * c^2}{d^2}$$

Where:

- a_x is the acceleration of a fundamental particle
- r_x is the classical radius of a fundamental particle
- c   is the speed of light
- d   is the distance between particles

Let's actually do the calculation:

$$a_x = 2.8179 \times 10^{-15} * c^2 = 253.26 \ m/s^2$$

Now we'll make a slight modification to Equation 46 so that it will pertain to an electron with the Bohr radius. We can do this very simply by dividing the classical radius of an electron by the fine-structure constant squared, (α^2), and multiply the speed of light squared also by the fine-structure constant squared.

Equation 47 – Acceleration by Charge

$$a_x = \frac{r_x}{\alpha^2} * c^2 * \alpha^2$$

Where:

a_x is the acceleration of the particle
r_x is the classical electron radius
c   is the speed of light
α is the fine-structure constant

Dividing the classical electron radius by the fine-structure constant squared, (α^2), gives us the Bohr radius, (5.2918×10^{-11}). Multiplying the speed of light by the fine-structure constant, (α), gives us the velocity of an electron wave. Some would claim that this is the velocity of an electron in its orbit. That is not quite accurate. That is the velocity of the electron wave. The electron is the whole sphere in which the electron wave is engulfed. The electron is not a particle in orbit, it's the whole package.

So now we can rewrite the equation again as follows:

Equation 48 – Acceleration by Charge

$$a_x = r_b * v^2$$

Where:

a_x is the acceleration of the particle
r_b is the Bohr radius – 5.2918×10^{-11}
v   is the velocity of an electron wave – 2.1877×10^6

The speed of an electron wave is:
$c * \alpha = 299{,}792{,}458 * 7.2974\times10^{-3} = 2.1877\times10^6$ meters/sec.

Now let's calculate the acceleration of an electron due to another charged particle one meter away:

$$a = 5.2918\times10^{-11} * (2.1877\times10^6)^2 = 253.26 \ m/s^2$$

As we can see from the calculations that this acceleration is equal to the acceleration calculated in Equation 46. So, the question is why two electrons with two different radii? The Bohr radius is the radius that we get from scientific observations. The classical electron radius is a theoretical radius that works out very well in scientific calculations. But what is the reality here? It appears the classical radius is the radius of an electromagnetic wave.

The reality is that electrons with the classical radius are apparently nowhere to be found. However, the mathematics for this electron works out very nicely for some reason, and that's a clue. The proposition being made here is that the photons of an electromagnetic wave and electrons are the same thing in different forms. The hypothesis is that electrons are created from photons. A photon locked into a circular orbit becomes an electron. We have already seen examples where electrons are created from electromagnetic waves, like Sunlight. Also, electrons can come into existence at just under the speed of light, as energy is converted into mass.

Lorentz to the Rescue

Let's test this hypothesis with the Lorentz transformation. Remember the Lorentz transformation is what Einstein used in his field equations to take into account the effects of relativity when determining the speed of an object. We can also use this transformation to calculate the radius of objects and take into consideration the effects of relativity. We'll use the Lorentz transformation on the classical electron radius to see what happens to the radius as we drop just under the speed of light. The equation for the Lorentz transformation follows:

Equation 49 – Transformation of Electron Radius

$$U_r = \sqrt{\frac{R_e}{1 - \frac{v^2}{c^2}}}$$

Where:
- R_e is the classical radius of an electron
- U_r is the unknown radius
- v is some velocity just under the speed of light
- c is the speed of light

Let's select a velocity that is very close to, but just under the speed of light. The speed of light is 299,792,458 meter/sec. An educated guess for the speed at which electrons would be created is 299,792,457.574938. Actually this was a predetermined value based on my own calculations. This number is just 0.425062 meters per second short of light speed, and is just 99.999999858214 percent of the speed of light. This calculation requires a high degree of accuracy and will not work on a simple calculator. Try doing this calculation on a computer. Now let's calculate:

$$5.29177\text{x}10^{-11} = \sqrt{1 - \frac{\frac{2.81794\text{x}10^{-15}}{(299792457.574938)^2}}{(299792458)^2}}$$

The unknown radius turns out to be the Bohr radius. Yes, the Bohr radius and the classical electron radius are exactly the same when seen through eyes of the Lorentz transformation. Electrons with the classical radius do exist at the edge of the universe but only for an instant in time before they are transformed into real Bohr radius electrons. This is where matter is created – at the top of the universe. This gives powerful support to the proposition that electrons are created from photons at just under the speed of light. It's at the top of the universe where all the action is taking place!

This also solves the mystery, and explains the origin of the fine-structure constant. The fine-structure constant is a relativistic relationship. Let's evaluate the expression for the Lorentz transformation:

$$5.3251\text{x}10^{-5} = \sqrt{1 - \frac{(299792457.574938)^2}{(299792458)^2}}$$

$5.3251\text{x}10^{-5}$ is equal to the fine-structure constant squared - α^2.
$5.3251\text{x}10^{-5} = (7.2973\text{x}10^{-3})^2$
$7.2973\text{x}10^{-3}$ is the fine-structure constant – α.

So when we divided the classical electron radius by the fine-structure constant squared, we were actually performing the Lorentz transformation without realizing it. Now we can understand why the fine-structure constant has been so useful to physicists. It captures the relativistic relationship in a concise form.

This transformation applies to protons as well as electrons. For protons we would calculate the Bohr radius of a proton by using the classical proton radius of 1.5347×10^{-18}. Once this transformation is made, what was a photon of an electromagnetic wave has been transformed into a fundamental particle of mass. Physicists have measured the Bohr radius quite precisely. Because of the nature of relativity, all the atoms in the universe will have the same properties and work the same way. It's a beautiful thing!

Creating Matter from Energy

Now let's take a closer look at the transformation of energy into matter, and paint the picture in words that will help us understand what actually takes place during this transformation. We are talking about the process of converting electromagnetic waves into matter. We are talking about what's happening at the top of the universe right now.

Electromagnetic waves actually consist of two intertwined waves: a negative electric wave and a positive electric wave. These waves rotate around each other as they travel through space. They both have wavelengths of the same size, although their amplitudes may be quite different. Because they are rotating, we could say that the amplitude is also the radius of the wave. Let's say that the radius of the negative electric wave is the classical electron radius, which is 2.8179×10^{-15} meters. Let's also say that the radius of the positive electric wave is denser and has a much smaller radius of 1.5347×10^{-18} meters. This is the classical radius of the proton and is about 1836.15 times smaller than the amplitude of the electron wave.

The universe is expanding in all directions at the maximum speed limit – the speed of light. Electromagnetic waves are also traveling at the same speed and direction as the expansion. But space itself is also expanding, attempting to grow outside the bounds of the universe. The catalyst for the creation of matter is the expansion of space. Unable to exceed the speed of light, the electromagnetic wavelengths at the very boundary of the physical universe are compressed to zero. As the wavelength compression is taking place, the electromagnetic waves begin to rotate faster attempting to maintain light speed. The amplitude of the wave increases starting at the classical radius and ending at the Bohr radius. When the wavelength reaches zero, just .425062 meters per second under of the speed of light, the rotational velocity has achieved its maximum speed and the amplitude of the wave has peaked at the Bohr radius. A transformation has taken place. The transformation of an electromagnetic wave into matter is illustrated in Figure 56.

Figure 56 – Wave Transformation

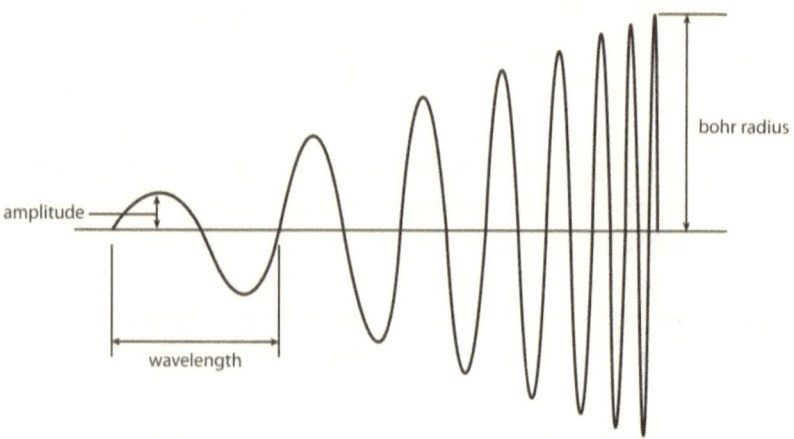

Courtesy of Brandon Fall

At that point the electromagnetic wave no longer exists. The negative electric wave still exists, and the positive electric wave still exists, but no longer as an electromagnetic wave. They have become separated. Both waves become locked into synchronous circular orbits, but no longer around each other. Each wave has taken on a separate identity; they have become individual entities. What was once a negative electric wave has become an electron; what was once a positive electric wave has become a proton. This is why matter has wave-like properties. Together they comprise a hydrogen atom. The radius of the electron is the amplitude of the electric wave. The radius of the proton is the amplitude of the proton wave. Electromagnetic energy has become matter, which is nothing more than electromagnetic waves divided into distinct pieces and locked into synchronous circular orbits.

Matter is just frozen energy. Once frozen in place this energy is now subject to all the laws of mass and motion. Our new particles begin to decelerate. All electrons are created at their lowest energy level; they have exactly the same size, and exactly the same amount of mass. Likewise, all protons have exactly the same size and exactly the same amount of mass. These are the only two particles created in this manner, and they still maintain their attraction for each other.

But what gives these newly created particles mass? Electromagnetic waves have no mass; they are pure energy and as such they have the freedom, perhaps the obligation, to travel through empty space at the speed of light. Even compressed into circular orbits the new particles are not physical entities; they have no substance. Neither electrons nor protons are physical particles. Once locked into circular orbits, their previous freedom of unhindered movement through space is denied. These newly created particles have now been endowed with the properties of inertia and momentum. The particles resist motion, and once in motion they resist any change to that motion. Inertia and momentum are the attributes that identify these new particles as mass, but there is no substance to mass. The amount of mass associated with a particle is determined by its measure of inertia and momentum. We can see that in Equation 46 – Acceleration by Charge.

Equation 46 shows that a particle's rate of acceleration, because of charge, has nothing to do with mass. The acceleration can only be attributed to the energy of the electromagnetic wave, the speed of light squared, and the radius of the particle. The smaller the radius, the greater the resistance to change in motion. It is the particles radius that determines its inertia.

Another attribute of these particles is electric charge, which is not something the particle acquires, but something that has always been present. Electromagnetic waves contain energy within themselves to propagate their own movement through space. When the wave is locked into a tiny circular orbit, that energy is also locked in place. This static energy is electric charge. Once separated, the two parts of an electromagnetic wave become a negatively charged electron and a positively charged proton. Though divided, they will forever maintain their attraction for each other. Combined, they make a hydrogen atom.

Even though these two new particles may be created from a variety of electromagnetic waves across the spectrum they are all created the same size. This size is determined not by the wavelength of the electromagnetic waves (which is reduced to zero) but by the amplitude. Judging from this, it would appear that the amplitudes of all electromagnetic waves are the same at the edge of the universe. This is why all electrons are exactly the same size, as are all protons. Whatever else the process of creation may be, it is exact and precise.

All matter originates from the same source - electromagnetic waves. Once electromagnetic waves are converted into mass they must obey all the laws pertaining to matter and motion. The newly created atoms of hydrogen consume space, which results in the attraction of other hydrogen atoms. As these atoms accumulate into ever larger structures they accelerate in the direction opposite the expansion. The rain of galaxies has begun.

Chapter 29

The Properties of Space

Space – The Final Frontier

The Permittivity of Space

In the latter half of the 19th century James Clerk Maxwell conducted experiments on the oscillating waves of both electric and magnetic fields. In order to help quantify the properties of these fields he developed the concepts of permittivity and permeability.

Permittivity, referenced by the Greek letter epsilon 'ε' or lower case 'e', is related to electric fields, and is a measure of farads per meter. A farad is a measure of capacitance (electric charge) measured in coulombs per volt. A coulomb is a measure of electrical charge expressed in terms of ampere seconds. A volt is a measure in difference of electrical potential given in terms of watts per ampere. A watt is a measure of power in terms of joules per second. A joule is a measure of energy given in terms of Newton meters. A Newton is a measure of force in terms of kilogram meters per second squared. Even after reading the above, most of us still don't know what permittivity is. We can break down permeability in the same manner to see the many levels of abstraction, or obfuscation, depending on how you look at it.

Each level of abstraction gives meaning and purpose to the units of measurement. Precise definitions and equations are available at each of the different levels to aid us in understanding and conceptualizing the laws of nature. But each step up, each layer of abstraction, is a step away from understanding the fundamentals of nature. In order to understand the origin of permittivity and permeability of space we must dig down to the lowest level possible. At its most basic units, permittivity is a measure of electric charge contained in a volume of space (kg/m^3). When permittivity is inverted, it can be understood as a volume of space per kilogram of electric charge.

The Permeability of Space

Permeability, referenced by the Greek letter mu "μ", is a measure of henrys per meter, and is related to magnetic fields. At its most basic units, permeability is a measure of meter-seconds^2 per kilogram (ms^2/kg). These kilograms are the kind that carry a magnetic field, or what we would call a magnetic charge. If permeability is inverted, it can be understood as a rate of change of kilograms of magnetic charge per meter per second squared.

The essence of permittivity and permeability is that they enable, or "permit," electromagnetic fields to flow through some medium, including empty space. The level of permittivity of a material determines how much of an electric field is enabled to flow through it. Also the level of permeability of a substance is a measure of its ability to enable the flow of a magnetic field.

The values of permittivity and permeability change depending on the material an electromagnetic wave is traveling through. For example, a magnetic field permeating through ferromagnetic materials like iron will be stronger than the magnetic field of air or other non ferromagnetic materials. Different materials have different capacities to hold electric and magnetic charges. In free space the values of permittivity and permeability are constant. The free space values are shown below. Note the subscript of "$_0$" on the symbol of reference is used to indicate free space values.

$$\text{Permittivity} \quad e_0 = 8.8542 \times 10^{-12} \quad \text{Farads/meter}$$
$$\text{Permeability} \quad \mu_0 = 4\pi \times 10^{-7} \quad \text{Henrys/meter}$$

Converting the units of measurement to their most basic form we would have:

$$\text{Permittivity} \quad e_0 = 8.8542 \times 10^{-12} \quad kg/m^3$$
$$\text{Permeability} \quad \mu_0 = 4\pi \times 10^{-7} \quad ms^2/kg$$

Multiplied together and inverted we would get the speed of light squared, and the units would be in meters squared per second squared. Equation 50 illustrates that relationship.

Equation 50 – Permittivity and Permeability

$$c^2 = \frac{1}{e_o \mu_o} \quad m^2/s^2$$

This calculation verifies the preceding statements:

$$c^2 = \frac{1}{8.8542 \times 10^{-12} * 4\pi \times 10^{-7}} = 8.98755 \times 10^{16} \quad m^2/s^2$$

Now we understand why the speed of light squared (c^2) is involved in the calculations used in the previous chapter to determine the acceleration of particles. The speed of light squared is directly related to the permittivity and permeability of space, which are related to electric and magnetic fields. In an electromagnetic wave there are actually two waves of opposite polarity: one electrically positive, the other electrically negative, both traveling and circling each other at the speed of light. That's why we use c^2 in the calculations. As the electromagnetic wave travels through space, it also generates a magnetic flux that surrounds it. This flux is brief and dissipates rapidly. The values for permittivity and permeability are the inverse of c^2. Curiously, these values are not divided equally.

Derivation of Coulomb's Constant

Coulomb's Constant can be thought of as a volume of space per kilogram of charge. Coulomb's Constant is derived from the permittivity of space as shown below:

Equation 51 – Coulomb's Constant

$$K_C = \frac{1}{4\pi e_o} = \frac{1}{4\pi\, 8.8542 \times 10^{-12}} = 8.98755 \times 10^9 \quad m^3/kg$$

To obtain the Coulomb Constant, the 4π was removed from permeability and applied to permittivity, then inverted. If we removed the 4π from permeability and inverted the value, we would get 1×10^7. The product of these two values is equal to the speed of light squared, as shown in the following calculation:

$$c^2 = K_c \ (8.98755\times10^9) \ast 1\times10^7 = 8.98755\times10^{16} \quad m^2/s^2$$

We have previously seen that the radius of a charged particle multiplied by the speed of light squared is somehow related to the acceleration of that particle. We also know that the inverse of the speed of light squared is equal to both the permittivity and the permeability of space. The constants for permittivity and permeability define the properties permeating space and surrounding each fundamental particle.

The Properties of Space

In attempting to understand the nature of gravity we became acquainted with something called gravitational flux, which is the contraction of space around a mass, and is defined by the mathematical expression $4\pi Gm$. Likewise, it's the permittivity of space around a mass that enables the electrical flux. The mathematical expression for electric flux is $4\pi e_o m$.

Permeability (μ_o) is defined as $4\pi\times10^{-7}$. Notice that 4π is already embedded in the value of μ_o. Taking the 4π from permeability and giving it to permittivity produces the same results when multiplying them together. $4\pi e_o$ and 10^{-7} multiplied together is equivalent to the expression $(e_o\mu_o)$. Henceforth we can refer to $4\pi e_o$ as the permittivity of space and 10^{-7} as the permeability of space.

The permeability of space around a mass enables the magnetic flux, and is defined by the expression $M_x\times10^7$. Notice that the permeability constant 10^{-7} is inverted to 10^7 in the expression for magnetic flux. The value calculated for the magnetic flux of an electron is: $9.1094\times10^{-31} \ast 10^7 = 9.1094\times10^{-24}$. This number is very close to the measured value listed by the NIST reference on constants for the magnetic moment of an electron: 9.2847×10^{-24}.[55] This confirms that our expression for magnetic flux is fairly accurate.

When we refer to magnetic flux and electric flux we are referring to the properties of space surrounding a fundamental particle that permit electric and magnetic waves to pass through that space. Permittivity enables electric flux and permeability enables magnetic flux. These fluxes "permit" the flow of electromagnetic waves through that space.

When Permittivity and Permeability are combined, the resulting value is the inverse of c^2. That value is 1.11265×10^{-17} $(s/m)^2$. This is the measure of the time it takes for an electromagnetic wave to travel through one meter of empty space, squared.

So far we have identified three fluxations of space: gravitational flux, electric flux, and magnetic flux. The word flux means to flow. In the case of gravity we have a flow of space. With electric flux we have a flow of an electric field, and with magnetic flux the flow is of a magnetic field. All three of these fluxations have the ability to impart acceleration to surrounding particles; thus, they all appear to possess a force.

Fields of Force

It's important to distinguish between the electromagnetic fields passing through the space occupied by particles of matter, and the waves that define the fundamental particles of matter. Niels Bohr showed us that electrons and protons reside in a stable state where they radiate no energy. At these stable states protons and electrons last forever. Remember that protons and electrons are waves locked in place. These frozen waves cannot radiate energy into the surrounding space or they would quickly dissipate. The energy of these waves is locked into the particles, so they cannot be the source of the surrounding electric and magnetic fields. However, the waves of these particles can and do interact with surrounding electromagnetic fields, and may be responsible for enabling them.

Where then do electric and magnetic fields come from? The magnetic and electric fluxations of space surrounding particles of matter induce electromagnetic waves to pass through the matter in a structured way. The structure of these waves then gives the particles the qualities of magnetic and electric charge. The details of these fluxations and the nature of charge will be discussed in detail in a later chapter.

Essentially there exists a constant flow of electric and magnetic waves through all particles of matter. The constant flow of these electric and magnetic waves is what we call electric and magnetic fields. All three fundamental particles of matter hold magnetic fields, but only electrons and protons hold electric fields. Neutrons are electrically neutral because they are composed of an electron and a proton whose electric fields negate each other. The magnetic fields of an electron and a proton combine to hold the neutron together and give the neutron a strong magnetic field.

It's the fluxations of space that gives us fields of force. Gravitational flux is generated as the result of the consumption of space; this is the "force" of gravity. Magnetic flux and electric flux is what creates the electric and magnetic fields surrounding particles of matter. We view these electric and magnetic fields as fields of force. Magnetic fields produce the magnetic charge and magnetism. Electric fields produce the electric charge and the flow of an electric field is electricity.

We will see shortly that electric and magnetic force fields are the result of fluxations of space. Particles of matter don't move on their own because of their surrounding electromagnetic fields. Particles move when their electromagnetic fields interact with the sum of all the electromagnetic fields from all the other particles surrounding it. Motion continues until equilibrium is found.

Chapter 30

The Electromagnetic Force

Fields of force can be electrifying.

Comparing Forces

Two of the most well known equations in physics are used to determine the force between objects. The first of these is Newton's Law of Gravitation, which is used to calculate the gravitational force between masses. The second is Coulomb's Law, used to calculate the force between electrically charged particles. Let's compare these two equations for a better understanding of the electromagnetic force.

Newton's Law   Coulomb's Law

$$F = \frac{Gm_1m_2}{d^2} \qquad\qquad F = \frac{K_cQ^2}{d^2}$$

Where:

F   is the force in Newtons
G   is the Gravitational Constant $= 6.6742\text{x}10^{-11}$ $m^3/kg\ s^2$
K_c is Coulomb's Constant $= 8.98755\text{x}10^9$ m^3/kg
Q^2 is electric charge squared $= 2.5670\text{x}10^{-38}$ kg^2/s^2
d   is the distance between the objects in meters.
m   is the mass of the objects.

Newton's gravitational equation has already been discussed. Coulomb's Law originates from French physicist Charles Augustin de Coulomb (1736-1806). It was Coulomb who first discovered and documented the relationship between charged particles. The calculation of force resulting from Coulomb's Law is often referred to as Coulomb's Force.

Let's compare the gravitational attraction between two electrons, with the force of electric charge between them. To simplify the calculations we'll set the distance between the two particles to one meter. The mass of an electron is 9.1094×10^{-31} kilograms. Using this mass with the values above we can calculate and compare the forces of each.

<table>
<tr><td align="center">Gravitational Force</td><td align="center">Coulomb's Force</td></tr>
<tr><td align="center">$F = \dfrac{Gm_1m_2}{d^2}$</td><td align="center">$F = \dfrac{K_c Q^2}{d^2}$</td></tr>
<tr><td align="center">$F = \dfrac{G * 9.1094 \times 10^{-31} * 9.1094 \times 10^{-31}}{1^2}$</td><td align="center">$F = \dfrac{8.9875 \times 10^{-11} * 2.567 \times 10^{-38}}{1^2}$</td></tr>
<tr><td align="center">$F = \quad 5.5383 \times 10^{-71}$ newtons</td><td align="center">$F = \quad 2.3071 \times 10^{-28}$ newtons</td></tr>
</table>

Our comparison shows us that Coulomb's Force is impressively stronger than gravity. The vast difference between these forces is significant. The force of electric charge, 2.3071×10^{-28} newtons, divided by the gravitational force, 5.5383×10^{-71} newtons, is 4.1657×10^{42}. The electromagnetic forces are extremely strong, much, much stronger than the gravitational force. This is why the gravitational effect is ignored when calculating interactions of electric charges.

Make a mental note of this difference in strength (4.1657×10^{42}), as we shall see it again and again. This difference in strength of force is so great we need to put it in perspective. A 10 penny nail is 3 inches long and weighs about ½ of an ounce. Drop the nail on the floor and pick it up with a small magnet. The electromagnetic force of the magnet has just overcome the entire gravitational attraction of the whole earth.

Coulomb's Constant

Related to Coulomb's Law is Coulomb's Constant, referenced by the letters K_c and is defined as 8.98755×10^9 $Nm^2/Coul^2$. In its most basic units of measurement it is defined as meters cubed per kilogram – m^3/Kg. This constant is analogous to the gravitational constant – G, and is to charged particles what the Gravitational Constant is to objects of mass. We have already seen in the last chapter how this constant was derived. This is essentially a volume of space per kilogram of charge. We'll now compare these two constants to better understand what they are.

The two columns below compare the Gravitational Constant with Coulomb's Constant. These columns show their values in the most basic units of measurement. They are both the inverse of density. The Gravitational Constant is a specific rate of volume consumption per kilogram, while Coulomb's Constant is simply a specific volume per kilogram of charge. The value for the Gravitational Constant is obtained by measurement, whereas the value of Coulomb's Constant is derived. In both these constants there is a volume of space per kilogram of matter.

For the Gravitational Constant the seconds squared in the denominator indicates a change of space per kilogram of mass. Coulomb's Constant is the amount of space required to contain the electrical charge of one kilogram of matter.

Gravitational Constant	Coulomb's Constant
6.6742×10^{-11}	8.98755×10^{9}
$\dfrac{\text{meters}^3}{\text{kilogram} * \text{second}^2}$	$\dfrac{\text{meters}^3}{\text{kilogram}}$

Just as the strength of gravity lies in the gravitational constant, so the strength of the electromagnetic force lies in Coulomb's Constant. Even with a casual comparison of these values one sees an unimaginable difference in strength. The difference between these two constants is 1.3466×10^{20}. This is another measurement of how much greater the electric force is than the gravitational force.

In the case of the gravitational constant we can interpret the units as the rate of contraction of space based on the amount of mass in kilograms. In the Space Consumption chapter we saw that the rate of space contraction, $4\pi G$, is a very small number indeed: 8.3870×10^{-10} cubic meters per kilogram of mass. This means that a large amount of mass actually consumes a relatively small amount of space. Thus gravity appears to be a very weak force.

As for Coulomb's Constant, there must be a different interpretation, even thought the units of measurement are similar. That understanding will be coming shortly. But, before attempting to explain Coulomb's Force, we need to understand another relationship.

Quantum Relationships

The product of the mass of a particle multiplied by its classical radius is identical for all fundamental particles. The mass and radius are also related to the elementary electric charge squared. The reason for the equality is easily explained. The mass of a proton is 1836.15 times larger than the mass of an electron, even though the proton's radius is 1836.15 times smaller than the electron's. Therefore the expression of mass multiplied by radius is the same for each fundamental particle. This is why the electric charges of protons and electrons are equivalent. The products of mass and radius are shown in the following calculations:

Particle:	Expression	Mass	*	Radius	$= Q^2 * 10^{-7}$
Electron:	Me Re	$9.1094e^{-31}$	*	$2.8179e^{-15}$	$= 2.567e^{-45}$
Proton:	Mp Rp	$1.6726e^{-27}$	*	$1.5347e^{-18}$	$= 2.567e^{-45}$
Neutron:	Mn Rn	$1.6749e^{-27}$	*	$1.5326e^{-18}$	$= 2.567e^{-45}$

Where:
  Me   is the mass of an electron
  Re   is the classical radius of an electron
  Mp   is the mass of a proton
  Rp   is the classical radius of a proton
  Mn   is the mass of a neutron
  Rn   is the classical radius of a neutron

Coulomb's Force

The unit of measurement used here for mass is a kilogram. Electric charge is referenced as Q. The unit of measurement used for electric charge is the coulomb which is measured in ampere seconds. One coulomb is the electric charge carried by a current of one ampere for one second. When Coulomb's Force is brought down to its most basic units of measurement it is measured in kilograms per second squared. But these are not ordinary kilograms, these kilograms are electrically charged particles.

Now that we've discussed the components, we can get back to analyzing two of the most widely used equations in physics. Let's compare the basic units of measurement in the two expressions shown below. These expressions come from the equations of force given at the beginning of the chapter. We'll assume a distance of one meter throughout this and the following chapters to eliminate unnecessary complexity in calculations. The following equations calculate the force between two particles.

Gravitational Force   Coulomb's Force

$$\frac{Gm_1m_2}{d^2} \qquad\qquad \frac{K_cQ^2}{d^2}$$

$$\frac{meters * kilogram}{second^2} \qquad\qquad \frac{meters * kilogram}{second^2}$$

The basic units of measurement for each expression are the same. Meters multiplied by kilograms per second squared is defined as a Newton, which is a measure of force. We have already seen that the force between charged particles is much greater than the force of gravity between the same two particles. We have already seen the big difference between the two constants, G and K_c. But there is also a big difference between the masses and the charges. The mass of two electrons multiplied together is 8.2981×10^{-61}. Two charges multiplied together equal 2.5669×10^{-38}. Although these are extremely small numbers, the difference is significant. The difference is 3.09345×10^{22}; this is a depiction of how much greater the magnetic force is than the gravitational.

This difference between masses and charges (3.09345×10^{22}) multiplied by the difference between the two constants G and K_c (1.3466×10^{20}) is equal to 4.1657×10^{42} newtons. This is the same difference in forces obtained at the beginning of this chapter.

Using Coulomb's force equation we can calculate the force between two charged particles one meter apart:

$$K_c Q^2 = 8.98755e^9 * 2.5669e^{-38} = 2.3071e^{-28}$$

Alternate Force Equation

Another equation for force that I learned about in high school is:
Force = Mass * Radius * c^2

This equation was not discussed in detail, probably because no one has a reasonable explanation as to how a mass, its radius, and the speed of light squared, can produce the electromagnetic force. But we, being the clever beings that we are, will deduce the essence of the electromagnetic force.

First let's validate this equation. The expression $M_x R_x c^2$ is equivalent to $K_c Q^2$. The following list of calculations using the expression $M_x R_x c^2$ shows that it gives the same results for each fundamental particle.

Table 32 – Electromagnetic Force

Particle:	Mass	*	Radius	*	c^2	= Force
Electron:	$9.1094e^{-31}$	*	$2.8179e^{-15}$	*	$8.9875e^{16}$	= $2.3071e^{-28}$
Proton:	$1.6726e^{-27}$	*	$1.5347e^{-18}$	*	$8.9875e^{16}$	= $2.3071e^{-28}$
Neutron:	$1.6749e^{-27}$	*	$1.5326e^{-18}$	*	$8.9875e^{16}$	= $2.3071e^{-28}$

The magnitude of force of electric charge for the fundamental particles is the same. The magnitude of that force can be calculated by multiplying the mass, the classical radius of the particle, and the speed of light squared. However, the neutron is composed of one proton and one electron, so the charges negate each other, leaving the neutron with no electrical charge. The magnitude of electric charge of the electron and proton is the same, only in opposite directions. The building blocks of all matter are made from these fundamental particles.

Let's closely examine these expressions to see how they are related:
The expression $K_c Q^2$ is equivalent to $M_x R_x c^2$.
$$K_c Q^2 = M_x R_x c^2$$

The value of c^2 is equal to $K_c * 1e^7$, so let's substitute $K_c * 1e^7$ for c^2.
$$K_c Q^2 = M_x R_x K_c 1e^7$$

Now let's factor out the K_c from both sides of the equation.

$$Q^2 \ = \ M_x\, R_x\, 1e^7$$

Now we know exactly where Q^2 comes from. It is not the product of two electric charges. It is the product of the mass, radius, and magnetic Flux ($1e^7$) of one particle. In Table 33 – Q^2, we can calculate the value of Q^2 for the fundamental particles:

Table 33 – Q^2

Particle	Mass	Radius	Flux		Q^2	Units
Electron	$9.10938e^{-31}$	$2.81794e^{-15}$	$1e^7$	=	$2.5669e^{-38}$	K_g^2/s^2
Proton	$1.67262e^{-27}$	$1.53470e^{-18}$	$1e^7$	=	$2.5669e^{-38}$	K_g^2/s^2
Neutron	$1.67493e^{-27}$	$1.53259e^{-18}$	$1e^7$	=	$2.5669e^{-38}$	K_g^2/s^2

The units of measurement for Q^2 is K_g^2/s^2; but that is NOT the square of a single charge. It is the product of Mass (K_g) and Magnetic Charge (K_g/s^2). Magnetic Charge is the product of the particle's Radius(m) * Magnetic Flux ($K_g\,/\,ms^2$), as shown in the following table. The resulting Units of Measurement for Magnet Charge is K_g/s^2.

Table 34 – Magnetic Charge, shows how the magnetic charge of each fundamental particle can be derived from the particles radius and the magnetic flux. More on this in the next chapter.

Table 34 – Magnetic Charge

Particle	Radius	Flux		Charge	Units
Electron	$2.81794e^{-15}$	$1e^7$	=	$2.81794e^{-8}$	K_g/s^2
Proton	$1.53470e^{-18}$	$1e^7$	=	$1.53470e^{-11}$	K_g/s^2
Neutron	$1.53259e^{-18}$	$1e^7$	=	$1.53259e^{-11}$	K_g/s^2

Coulomb's Force Transformed

Now that we have an alternate way of expressing the force between particles, let's make some minor adjustments to Coulomb's Force. Let's change the expression K_CQ^2 to its equivalent, $M_xR_xc^2$, as shown below. Also shown below are the units of measurement.

<u>Coulomb's Force</u>   <u>Coulomb Transformed</u>

$$\frac{K_CQ^2}{d^2}$$

$$\frac{M_xR_xc^2}{d^2}$$

$$\frac{meters * kilogram}{second^2}$$

$$\frac{meters * kilogram}{second^2}$$

The units of measurement remain the same, as well as the calculated values of these expressions. The two expressions are equivalent. Once again, we haven't changed anything other than our perception of the electromagnetic force. Now we don't need the value of an electric charge or a derived constant. We can now view the electromagnetic force in terms of the most fundamental particles and the speed of light squared.

Once again, it would help to understand Coulomb's Force in terms of acceleration rather than force. So we'll perform the same manipulation here as we did for gravity. The inertial force is equivalent to the electromotive force as shown in the equations below:

$$F = Ma \qquad\qquad F = \frac{M_xR_xc^2}{d^2}$$

By equating these two expressions of force we obtain the following equation:

$$Ma = \frac{M_xR_xc^2}{d^2}$$

By dividing each side of the equation by mass, we get an equation for the acceleration of a charged particle by Coulomb's force:

Equation 52 – Acceleration by Coulomb's Force

$$a_x = \frac{R_xc^2}{d^2}$$

Now we can calculate the acceleration of an electron and a proton due to the electromagnetic force between them. Once again we're using a distance of one meter to eliminate the division by distance squared and simplify the calculations. Shown below are the calculation for the acceleration of an electron and a proton by Coulomb's Force. The first is for the acceleration of an electron; the second for the acceleration of a proton.

$$a_e = \frac{R_e c^2}{d^2} = \frac{2.81794e^{-15} * 8.98755e^{16}}{1^2} = 253.26 \ \frac{m}{s^2}$$

$$a_p = \frac{R_p c^2}{d^2} = \frac{1.53472e^{-18} * 8.98755e^{16}}{1^2} = .13793 \ \frac{m}{s^2}$$

From these calculations we see that the acceleration of the fundamental particles is dependent on their classical radius and the speed of light squared. We can also see that their accelerations are proportionate to their radius. The classical radius of an electron is 1836.15 times larger than the classical radius of a proton. This means the acceleration of an electron is also 1836.15 times greater than that of the proton, as shown in the calculation below.

$$253.26 \div .13793 = 1836.15$$

Difference in Strength

Russian born scientist George Gamow (1904-1968) states in his book Gravity, "Any theory which claims to describe the relation between electromagnetism and gravity must explain why this electric interaction between two particles is approximately 10^{40} times larger than the gravitational interaction."[56] That explanation is about to come.

Let's once again compare the difference in strengths of the electromagnetic and gravitational forces. Only this time we'll make the comparison based on the accelerations of the fundamental particles. The equation for acceleration by electromagnetic force is given above. Once again let's dispense with d^2 in both expressions and assume a distance of one meter between the particles.

This calculation shows the ratio of accelerations of an *electron* due to the electromagnetic force and gravitational force.

$$\frac{R_e c^2}{G \, M_e} = \frac{2.8179e^{-15} * c^2}{G * 9.1094e^{-31}} = 4.1657e^{42}$$

We have just compared the accelerations of the fundamental particles due to electromagnetic and gravitational forces at a distance of one meter. Let's make the comparison again only this time let's not specify a distance for the accelerations. Rather than a difference of accelerations, we'll compare a difference of volumes. Let's add 4π to both the numerator and denominator to get the following equation:

$$\frac{4\pi R_e c^2}{4\pi G M_e} = \frac{m^3/s^2}{m^3/s^2}$$

The 4π in the numerator and denominator will cancel out of the calculations, but provides a better understanding of what we are actually comparing. In the denominator we now have the expression that gives us the gravitational flux, which we've seen in the chapters on gravity. This is the volume of space contracted by this particular particle. In the numerator we see the volume of space occupied by an electromagnetic wave. We are comparing rates of change in volumes of space. This comparison shows again how much stronger the electromagnetic force is than the gravitational force. It has to do with the size of space affected. The denominator shows us the small amount of space consumed by a fundamental particle of matter. The numerator shows us the fluxation of space occupied by the electromagnetic field surrounding that fundamental particle. The volume of space occupied by the electromagnetic field is much greater than the space consumed by the particle. In fact it's $4.1657e^{42}$ times greater.

We see that the electromagnetic force derives its energy and strength from electromagnetic waves traveling at the speed of light. This is why the speed of light squared is involved in these equations. But we still have not quite captured the essence of what's behind the electromagnetic force. There is more to come, and we still need answers to some basic questions. First we need to understand a little more about the nature of charge.

Chapter 31

The Nature of Charge

Life is full of charges.

The Three Charges of Life

There are three charges in all matter. First of all what is matter? All matter is made from Electrons, Protons, and Neutrons. Each of these are a form of matter. All mass in the universe is made from these three fundamental particles of matter. If something is not made from these – it is not matter. There is no such thing as "Dark Matter". All three of these fundamental particles possess the property of gravity. Having this property is what makes these fundamental particles of matter have mass.

In this chapter we will identify the three charges associated with mass. The three charges are Gravitic, Magnetic, and Electric. The values for each of these charges is specified in kilograms. The symbol used in physics to identify a kilogram of matter is Kg. So, normally when we see this symbol we associate it with a specific amount of mass. Mass can now also be referred to as Gravitic Charge. But, matter also possesses a specific amount of electric charge and magnetic charge. So the symbol Kg refers to Gravitic, Electric, and Magnetic charge, since all three charges are possessed by matter. Later we will see how they relate to each other.

In the Equations and calculations that follow we may use the indicated symbols to denote the Charge types:

Charge Types

Gravitic:	Q_1	Kg
Magnetic:	Q_2	Kg / s^2
Electric:	Q_3	Kg

Gravitic Charge

As part of understanding the nature of charge let's take a look at gravitational charge. To be consistent with the other charges, let's refer to this charge as Gravitic Charge. Some may suggest that there is no such thing as Gravitic Charge. But, considering gravity as an effect of gravitic charge isn't a new idea; I learned this in High School. This is just a means to help us understand the nature of charge, and how it compares with electric and magnetic charge.

Let's calculate the mass of an object as described in the chapter on Space Consumption. The space contraction equation in Equation 8 – The Law of Volumes divided by Equation 14 – Space Consumption Rate is equivalent to the mass. In fact this is how mass is defined and is key to understanding the electromagnetic force. Once again, we are dividing the amount of space consumed by a fundamental particle, by the rate of space consumption, giving the particles mass. The equation for Gravitic Charge follows:

Equation 53 – Gravitic Charge

$$Q_1 = \frac{4\pi r^2 a}{4\pi G} = \frac{r^2 a}{G} \quad Kg$$

Where:
    Q_1 - is the Gravitic Charge of a fundamental particle
                (also known as Mass)
    r   - is the distance from center: $2.46885e^{-13}$ for electron
    a   - is the acceleration – classical radius
    G   - is the Gravitational Constant – $6.6742e^{-11}$

Since the volume of space consumed by any mass is proportional to the size of the mass, we could, and have been calculating, the acceleration at any particular distance. In this case we want to select the distance at which the acceleration is the same as the particle's classical radius. Lets calculate the mass of an electron using this equation. The electron's classical radius is $2.81794e^{-15}$ meters, and the distance from center is $1.46885e^{-13}$ meters. (If necessary the reader may want to review the Chapter on Space Consumption to refresh their memory)

$$Q_1 = \frac{(1.46885e^{-13})^2 * 2.81794e^{-15}}{6.6742e^{-11}} = 9.10938e^{-31} \ \ Kg$$

The mass (Gravitic Charge) of a fundamental particle is the result of the consumption of space. In this case we have an electron.

$$Me = 9.10938e^{-31} \quad Kg$$

Now let's use the same equation to calculate the mass of a Proton. The proton's classical radius is $1.53470e^{-18}$ meters, and the distance from center is $2.69704e^{-10}$ meters.

$$Mp = \frac{(2.69704e^{-10})^2 * 1.53470e^{-18}}{6.6742e^{-11}} = 1.67262e^{-27} \ \ Kg$$

The mass (Gravitic Charge) of a Proton is:

$$Mp = 1.67262e^{-27} \quad Kg$$

Gravitic Charge is equal to the volume of space being consumed ($4\pi r^2 a$) divided by the rate of consumption ($4\pi G$). There's only one kind of gravitic charge, and it is one of attraction.

It is unimportant whether we consider the gravitational attraction between two masses as the attraction of the masses or the attraction of gravitic charge. They are the same thing. The term gravitic charge could have easily been used to define the attraction between all objects instead of something called mass. The term gravitic charge is synonymous with mass. There is a good reason for thinking in terms of gravitic charge rather than mass, and we shall investigate this shortly.

Before going on to Electric and Magnetic Charges let's look at the difference between the mass of an electron and the mass of a proton.

$$\frac{Mp = 1.67262e^{-27} \quad Kg}{Me = 9.10938e^{-31} \quad Kg}$$

$$Diff = 1,836.15$$

Here we see the mass of a proton is considerable larger, 1,836.15 times larger, than an electron, even though the electron radius is larger than the proton radius. The electron radius is in fact 1,836.15 times larger than the proton radius.

Magnetic Charge

We saw in the previous chapter that Magnetic Charge was the product of the classical radius of a fundamental particle and Magnetic Flux ($1e^7$). The Equation for Magnetic Charge is shown in Equation 54.

Equation 54 – Magnetic Charge

$$Q_2 = R_x * 1e^7 \; \frac{Kg}{s^2}$$

Where:
  Q_2 - is the Magnetic Charge of a fundamental particle
  R_x - is the classical radius of a fundamental particle
  $1e^7$ - is the Magnetic Flux of Space (Permeability)

The calculations below show the magnetic charges of the fundamental particles as shown in Table 34 – Magnetic Charge, in the previous chapter. The value for magnetic charge is specified in kilograms per seconds squared – Kg/s^2. These kilograms do not refer to kilograms of mass, but kilograms of magnetic charge. The s^2 in the denominator implies that magnetic charge is an accelerating flow of magnetic flux.

Let's look at the difference between the magnetic charge of an electron and the magnetic charge of a proton.

$$\frac{Q_e = \quad 2.8179e^{-8} \quad Kg/s^2}{Q_p = \quad 1.5347e^{-11} \quad Kg/s^2}$$

Diff =   1,836.15

Here we see that the magnetic charge of an electron is 1,836.15 times larger than the magnetic charge of a proton. Remember that the difference of the gravitic charges of an electron and proton also differ by 1,836.15. Now let's multiply the gravitic and magnetic charges of each particle to see how they balance:

	Gravitic Charge	*	Magnetic Charge	=	Q_1 * Q_2
Electron	$9.10938e^{-31}$	*	$2.81794e^{-8}$	=	$2.56697e^{-38}$
Proton	$1.67262e^{-27}$	*	$1.53470e^{-11}$	=	$2.56697e^{-38}$
Difference	- 1,836.15		+ 1,836.15		

Here we can see that the differences in magnitude balance out. So that when an electrons gravitic charge is combined with its magnetic charge, the value is the same as the protons gravitic and magnetic charges combined. Thus, the value of the combined gravitic and magnetic charges are the same for electrons and protons. This is also true of neutrons as shown in Table 33. The math for neutrons was dispensed with in order to prevent the reader from dying of boredom.

The value of the combined charges is given in units of measurement of kilograms squared divided by seconds squared – Kg^2/s^2. However, the charges of (Kg) for each is of a different kind. So what we have is a gravitic charge combined with a accelerating magnetic charge. There's nothing happening here – yet.

An explanation of how magnetic charges are created and what they do will be discussed in detail in a subsequent chapter.

Electric Charge

The Electric Charge does not stand on its own. The Electric Charge must be combined with the Magnetic Charge because they share the same space; specifically the classical radius of a fundamental particle. For Magnetic Charge we divided the particles classical radius by the Permeability of space Constant. For Electromagnetic Charge we divide the particles classical radius by the Permeability of space Constant, and the Permittivity of space Constant, and the Gravitic Constant. The Electromagnetic Charge results from a combination of all three properties of space. For simplicity sake, we often refer to the Electromagnetic Charge as the Electric Charge.

Equation 55 – Electromagnetic Charge

$$Q_3 = \frac{R_x}{4\pi e_o \, 1e^{-7} \, G} \quad K_g$$

Where:

Q_3 — is the Electromagnetic Charge of a fundamental particle
R_x — is the classical radius of a fundamental particle
G — is the Gravitational Constant $6.6742e^{-11}$
e_o — is the Permittivity of Space $8.8542e^{-12}$
$1e^{-7}$ — is the Permeability of Space

$$Q_3 = \frac{2.81794e^{-15}}{4\pi * 8.8542e^{-12} * 1e^{-7} \, 6.6742e^{-11}} = 3.79467e^{12} \quad K_g$$

$$Q_3 = 3.79467e^{12} \quad K_g$$

How do we know this value is correct? Divide this charge by gravitational charge ($3.79467e^{12}$ / $9.10938e^{-31}$ = $4.16567e^{42}$) and the difference is right on target.

Combining Charges

In this chapter we have examined three kinds of charges; Gravitic, Magnetic, and Electromagnetic. All three charges are essential components of electrons and protons. Charles Coulomb was the first to capture all the essential information to formulate the electromagnetic force equation. The following is Coulomb's Force equation:

$$F_e = \frac{K_c Q^2}{d^2} \quad \frac{Kg \, m}{s^2}$$

In the equation for Coulomb's Force, Q^2 represents the value of the charges of two electrons, also referred to as an electric or electron charge. The magnitude of the electric charge squared is $2.56697e^{-38}$. Coulomb viewed these two charges as being from two different electrons with charges of equal value $-1.60217e^{-19}$.

Coulomb no doubt compared his equation with Newton's equation for the force of gravitational attraction. We can see why Coulomb envisioned two separate electric charges in his force equation. But, now we have discovered that these are not two different electric charges from two different particles; They are two different kinds of charges from the same particle. The magnetic charge and the gravitic charge combined make Q^2. Splitting these charges equally in half makes no sense. That is why there is no mathematical use for a split charge. So now let's transform Coulomb's equation using the charges that have been identified in the preceding sections.

Equation 56 – Electromagnetic Force

$$F_x = \frac{K_c \quad * \quad Q_1 \quad * \quad Q_2}{d^2} \qquad \frac{Kg\, m}{s^2}$$

Where:

F_x	is the Electromagnetic Force
K_c	is Coulomb's Constant
Q_1	is the Gravitic Charge of a Fundamental Particle
Q_2	is the Magnetic Charge of a Fundamental Particle

The only change that was made to Coulomb's Force equation was to replace the Q^2 with the Gravitic Charge (Q_1), and the Magnetic Charge(Q_2). Now let's value the equation to calculate the value of the electromagnetic force of an electron.

$$F_e = \frac{8.98755e^9 * 9.10938e^{-31} * 2.81794e^{-8}}{1^2} \qquad \frac{Kg\, m}{s^2}$$

$$F_e = 2.3071e^{-28} \qquad \frac{Kg\, m}{s^2}$$

As we can see from the results, the electromagnetic force is a combination of Coulomb's Constant, the Gravitic and Magnetic Charges. It takes all three items combined and working together to make the electromagnetic force. But what happened to the Electromagnetic Charge. We shall see shortly.

The following table shows the three Charges of the three Fundamental Particles.

	Constant	Electron Charge	Proton Charge	Neutron Charge
Gravitic:	Q_1	$9.10938e^{-31}$	$1.67262e^{-27}$	$1.67493e^{-27}$
Magnetic:	Q_2	$2.81794e^{-8}$	$1.53470e^{-11}$	$1.53259e^{-11}$
Electric:	Q_3	$3.79467e^{12}$	$2.06664e^{9}$	0

The combined Gravitic and Magnetic Charges of each Fundamental Particle is equal to $2.56697e^{-38}$ Kg^2/s^2; as shown in the calculations below.

	Constant	Electron Charge	Proton Charge	Neutron Charge
Gravitic:	Q_1	$9.10938e^{-31}$	$1.67262e^{-27}$	$1.67493e^{-27}$
Magnetic:	Q_2	$2.81794e^{-8}$	$1.53470e^{-11}$	$1.53259e^{-11}$
Product:	$Q_1 * Q_2$	$2.56697e^{-38}$	$2.56697e^{-38}$	$2.56697e^{-38}$

Chapter 32

The Unification of Forces

"The day when we shall know exactly what electricity is will chronicle an event probably greater more important, than any other recorded in the history of the human race." – Nikola Tesla

The Electromagnetic Force

Throughout the last few chapters we've studied, revised and compared many equations to increase our understanding of the forces of nature. This has been a long and laborious process. Our efforts will not go unrewarded.

Since the days of Michael Faraday scientists have been searching for the connection between the forces of nature. That connection has proven elusive, remaining hidden behind a shroud of mystery. The unification of forces was unattainable – until now! That shroud could not be removed until the nature of space had been realized. Now we can see and understand the dynamics of space and its affect on the characteristics of that space.

Where is the most obvious place to look for the unification of forces? Is it not in the heart of the most fundamental particles of matter? The fluxations of space converge within the fundamental particles of matter, and are the source of all forces and motion of matter. The convergence of these fluxations is where the unification of these forces occurs.

Equation 56 – Electromagnetic Force, gives the value of the electromagnetic force of charged fundamental particles. That, of course, would be the electron and the proton. The neutron goes through the same process, but since it contains an electron and a proton it is electrically neutral. However, the magnetic fields of each combine to yield a magnetic charge.

Calculating Forces

The table below shows the equations used to calculate the Forces of nature. These are defined as the 'Old Equations'. In order to get a better understanding of how these Forces are related, we have revised the Old Equations to New Equations which show their common components. The resulting calculations still yield the same results, but now, in this format, we can see how similar these three Forces of Nature are. They are all a product of the Gravitic Constant, Mass, and the associated Charge.

Type of Force	Old Equation		New Equation		Force
Gravitic	$G\,M_1M_2$	→	$G\,M\,Q_1$	=	$5.53831e^{-71}$
Magnetic	$G\,Q_1Q_2$	→	$G\,M\,Q_2$	=	$1.71325e^{-48}$
Electric	$K_c\,Q_1Q_2$	→	$G\,M\,Q_3$	=	$2.30708e^{-28}$

The following sections will show how these three equations are actually rolled up into one equation.

Unified Forces

Equation 56 is repeated here for convenience, as we will be breaking it down into its most basic components, and then transforming it into an equation that shows all forces are united. Then we will be able to see that the electromotive force of the fundamental particles is made from the three charges described in detail in the previous chapter.

$$F_x = \frac{K_c \quad * \quad Q_1 \quad * \quad Q_2}{d^2} \qquad \frac{Kg\,m}{s^2}$$

In the following transformations we will assume a distance of one meter and eliminate the d^2 from the denominator of the equation. Coulomb's Constant will be transformed into its equivalent value $1/4\pi e_o$. For the transformation of Gravitic Charge (Q_1) see Equation 53 – Gravitic

Charge, in the previous chapter. For the transformation of Magnetic Charge (Q_2) see Equation 54 – Magnetic Charge. Using this equation we will be calculating the Electromagnetic force of an electron. Remember that we have removed the 4π from the permeability constant and redefined μ_0 to be $1e^{-7}$.

Equation 57 – United Forces

$$F_x = \frac{1}{4\pi e_o} * \frac{R_x\, r^2}{G} * \frac{R_x}{1e^{-7}} \qquad \frac{Kg\, m^3}{s^2}$$

Where:

F_x	is the Electromagnetic Force
R_x	is the classical radius of the fundamental particle
r	is the distance from center - for electron $1.46885e^{-13}$
e_o	is the permittivity of space - $8.8542e^{-12}$
$1e^{-7}$	is the permeability of space
G	is the Gravitational Constant - $6.6742e^{-11}$

In the numerator of Equation 57, we have nothing but space. In the denominator we have all the properties of space. What we have here is the unification of forces all in one simple equation. The force of every fundamental particle contains gravitic, magnetic, and electric forces. They are all here in every particle. All we need to do now is break the equation apart into its constituent components and analyze them separately. Surely, a simple task.

But, before we start our analysis, let's validate this equation for the electron. The classical electron radius is $2.8179e^{-15}$, and the radius, (distance from center) is $1.46885e^{-13}$. The expression $4\pi e_o$ is equal to $1.11265e^{-10}$.

$$F_e = \frac{1}{1.11265e^{-10}} * \frac{6.07978e^{-41}}{6.6742e^{-11}} * \frac{2.81794e^{-15}}{1e^{-7}} \qquad \frac{Kg\, m^3}{s^2}$$

$$F_e = 8.98755e^9 * 9.1094e^{-31} * 2.81794e^{-8}$$

$$F_e = 2.3071e^{-28}$$

The equation just used to calculate the force of an electron was Coulomb's Electromagnetic Force Equation. We multiplied Coulomb's Constant with the mass of an electron (Gravitic Charge), and the Magnetic Charge. This gave us the correct results, but what happened to the Electric Charge? How is it that we can multiply Gravitic Charge and Magnetic Charge and get Electromagnetic Force? This form of the equation can be misleading. Let us now rewrite Coulomb's Force Equation in a different way; as was presented in a previous section.

Instead of Coulomb's Force Equation:

$$F_x = \qquad K_c \qquad * \qquad Q_1 \qquad * \qquad Q_2$$

Let's use this new equation:

$$F_x = \qquad G \qquad * \qquad Q_1 \qquad * \qquad Q_3$$

This is essentially the same equation as Coulomb's Force equation. However, in this equation we replaced Coulomb's Constant (Kc) with the Gravitic Constant (G), and we replaced Magnetic Charge (Q_2) with Electromagnetic Charge (Q_3). This will be a more effective way of looking at the Electromagnetic Force equation, and it gives us the exact same results. Once again we need to transform Gravitic Charge (Q_1) and Electric Charge (Q_3) to its more basic form as shown in the previous chapter. And again we will be calculating the Electromagnetic force of an electron. Remember that we have redefined μ_o to be $1e^{-7}$.

Equation 58 – United Gravitic Forces

$$F_x = G * \frac{R_x r^2}{G} * \frac{R_x}{4\pi e_o \, 1e^{-7} \, G} \qquad \frac{Kg \, m^3}{s^2}$$

Where:

F_x	is the Electromagnetic Force
R_x	is the classical radius of the fundamental particle
r	is the distance from center - for electron $1.46885e^{-13}$
e_o	is the permittivity of space - $8.8542e^{-12}$
$1e^{-7}$	is the permeability of space
G	is the Gravitational Constant - $6.6742e^{-11}$

Now let's once again validate this equation for an electron:

$$F_e = 6.6742e^{-11} * \frac{6.07978e^{-41}}{6.6742e^{-11}} * \frac{2.81794e^{-15}}{4\pi e_o \ 1e^{-7} \ G} \ \frac{Kg \ m^3}{s^2}$$

$$F_e = 6.6742e^{-11} * 9.1094e^{-31} * 3.79467e^{-12}$$

$$F_e = 2.3071e^{-28}$$

Same equation – same results.

Electromagnetic Charge

The following expression is the Electromagnetic Charge. But, also notice that the Magnetic Charge is embedded within the Electromagnetic Charge. I suspect that's why they called the whole expression the Electromagnetic Charge. So when we calculate the Electromagnetic Force, we use the Gravitic Constant (G), and the Gravitic Charge (Q_1), and the Electromagnetic Charge (Q_3), which also includes the Magnetic Charge (Q_2). It takes all three Charges combined to create the Electromagnetic Force.

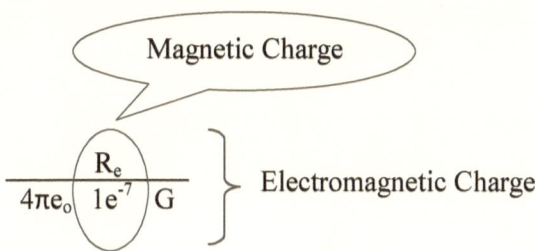

The Table below shows the three forces generated by the three charges of an electron. All three forces are operating at the same time within the fundamental particles. Notice the difference in magnitude between the Electric Charge and the Gravitic Charge – $4.16567e^{42}$. That is the same difference seen in previous chapters and the same difference seen here in the Force comparison in the table below.

	Gravitic Constant	Electron Mass	Electron Charge		Force	
Gravitic:	G	$9.10938e^{-31}$	$9.10938e^{-31}$	=	$5.53831e^{-71}$	Kg m³/s²
Magnetic:	G	$9.10938e^{-31}$	$2.81794e^{-8}$	=	$1.71325e^{-48}$	Kg m³/s⁴
Electric:	G	$9.10938e^{-31}$	$3.79467e^{+12}$	=	$2.30708e^{-28}$	Kg m³/s²
Diff E/G			$4.16567e^{42}$		$4.16567e^{42}$	

One very important consideration we can glean from these calculations is that we can now see that Gravity plays a very important role in the generation of all three Fundamental Forces, Gravitic, Electric and Magnetic.

Chapter 33

Space Transformations

A change of space will do you good!

Space Transformations

We have just seen the powerful electromagnetic force created from the three charges combined. Where is this force really coming from? It can't just appear out of empty space. Or can it? Where does the energy come from to power this force? The mysterious electromagnetic force has an origin. The electromagnetic force has a source that is about to become obvious.

In the chapter on Space Consumption we saw how each fundamental particle contracts space to itself. It was stated emphatically that the space was consumed out of existence. That is true but, there's more to it than just that. A transformation takes place – a Space Transformation. The space consumed by the fundamental particles has its properties converted into electric and magnetic charges.

We have a volume of space being contracted to all the fundamental particles with nothing happening until the associated matter consumes the space. The space, with the aid of its properties, is transformed into electric and magnetic charges. The energy that was dormant in the consumed space, is reorganize and released in a structured manor and is called electric and magnetic charge.

We know mass has energy locked within. Any equations showing the energy of mass would be showing it in a dormant state, just like the energy contained in space is in a dormant state. Once that energy in mass is released, it takes on a different life. So it is with the dormant energy held within space. That energy is released when a transformation occurs to bring it to life.

We can calculate the energy that powers the forces by dividing the force by the classical radius of the fundamental particle. Table 35 shows the energy generated from space for the Gravitic, Magnetic, and Electric Forces:

Table 35 – Electromagnetic Energy

	Electron Force		Electron Radius		Electron Energy	
Gravitic:	$5.53831e^{-71}$	÷	$2.81794e^{-15}$	=	$1.96537e^{-56}$	Kg m^2/s^2
Magnetic:	$1.71325e^{-48}$	÷	$2.81794e^{-15}$	=	$6.07978e^{-34}$	Kg m^2/s^4
Electric:	$2.30708e^{-28}$	÷	$2.81794e^{-15}$	=	$8.18710e^{-14}$	Kg m^2/s^2

Electromagnetic Energy

Another minor modification to Equation 58 – United Gravitic Forces, is to factor out the Gravitic Constant (G) from the numerator and the denominator. Also, we can eliminate (divide out) the classical radius of the fundamental particle. The resulting equation gives the energy of the Electromagnetic Charge.

Equation 59 – Electromagnetic Energy

$$E_x = \frac{R_x r^2}{4\pi e_o \, 1e^{-7} \, G} \quad \frac{Kg\ m^2}{s^2}$$

Where:

- E_x is the Electromagnetic Energy
- R_x is the classical radius of the fundamental particle
- r is the distance from center - for electron $1.46885e^{-13}$
- e_o is the permittivity of space   - $8.8542e^{-12}$
- $1e^{-7}$ is the permeability of space
- G is the Gravitational Constant   - $6.6742e^{-11}$

Let's evaluate this equation to calculate the energy created by an electron from 'empty' space.

$$E_x = \frac{6.07978e^{-41}}{1.11265e^{-10} * 1e^{-7} * 6.6742e^{-11}} \quad \frac{Kg\ m^2}{s^2}$$

$$E_x = 8.1871e^{-14}$$

This is exactly the same amount of energy as possessed by an electron. Remember Einstein's equation: $E = mc^2$.

$$E = 9.10938e^{-31} * 8.98755e^{16} = 8.1871e^{-14}$$

Energy From Space

So, how did we get all this energy from empty space? The answer is that space contains dormant energy. The fundamental particles are capable of converting the dormant energy in empty space into the dynamic energy of Charge.

Now let's isolate an expression from Equation 59. The following expression uses the same space that is consumed, the space is divided (processed) by the permeability of space constant ($1e^{-7}$). Dividing space by the permeability of space constant, converts 'empty' space into energy.

$$E_x = \frac{R_x r^2}{1e^{-7}} = \frac{6.07978e^{-41}}{1e^{-7}} = 6.07978e^{-34} \ \frac{Kg\,m^2}{s^4}$$

This is the same volume of space that is being consumed, compare with Table 35. But, as the space is consumed, it is converted into magnetic energy. So we can clearly see now the functioning of these properties of space. The Gravitic Constant tells us the amount of space being consumed per kilogram of matter. The permeability of space enables the conversion of space into the energy required for the Magnetic Charge, As the space is consumed, it is converted into energy.

This energy is enhanced by the processing of the Gravitic Constant and the Permittivity of Space ($4\pi e_o$) to create all the energy required for the Electromagnetic Charge. All three of the properties of space are used to transform space into energy for an Electromagnetic Charge. As the space is consumed, it is converted into energy.

The following expression illustrates the completion of this process. Equation 59 gives the solution to the whole process.

$$E_x = \frac{E_m}{4\pi e_o G} = \frac{6.07978e^{-34}}{7.42605e^{-21}} = 8.1871e^{-14} \ \frac{Kg\,m^2}{s^2}$$

The Energy of Space

In the previous calculations we see that incredibly powerful forces are created from within the heart of the fundamental particles. Space is constantly being contracted toward mass. Right between the Bohr radius and the classical radius of fundamental particles, is the Compton Radius. This is where a certain amount of space is squeezed, squished, or squashed out of existence. Which is a better technical term, squeezed, squished, or squashed? Anyway, as space is squished out of existence, that space is transformed into electromagnetic fields. Electromagnetic fields are created, because the properties of permeability and permittivity exist in space. The process starts with mass consuming space creating gravity. Gravity is the key to the whole process. The properties within the consumed space enable the transformation of that space into electromagnetic energy fields.

This concept of creating energy from space is the essence of the forces of nature; and needs to be repeated for emphasis and clarity. So let's take another look at the energy of space and the creation of the electromagnetic charges from space. The mathematical expression below combines the three properties of space: gravity, permittivity and permeability. When all three of these properties are used to process a small volume of space an incredible transformation takes place. The expression below shows the volume of space consumed, divided by (processed by) the properties of space. The latent energy of that space is transformed into energy for the electric and magnetic charges.

$$\frac{\text{space consumed}}{G \, e_o \, \mu_o} \qquad \frac{m^3 / s^2}{m / Kg}$$

$$\frac{6.07978e^{-41}}{7.42605e^{-28}} \quad = \quad 8.1871e^{-14} \quad \frac{Kg \, m^2}{s^2}$$

In this process the space is consumed. The properties of that space are used to energize the electric and magnetic charges. The space, without its properties will collapse out of existence. This results in the gravitational attraction that mass possesses. The electric charge is generated from the permittivity of space, and the magnetic charge is generated from the permeability of space. So all three of the properties of space work together to produce the forces of nature. These forces create the world in which we live.

So now we have the solution to the mystery of these forces. The electromagnetic force is generated by the Space Transformation from the properties of space into an electromagnetic energy field. The energy released is equivalent to $E = mc^2$. Take a deep breath! This has been the quest of many brilliant scientists who would have loved to discover this elusive connection of the forces of nature. Yet it was here all the time, very simple, but never clearly understood. Maybe it's time for a good cold brew. But, before imbibing, let's continue, we are not quite finished!

Energy Density of Space

In the previous section we indicated how much energy was released by showing the energy contained within a proton and an electron. This may be true, but that energy didn't come from an electron or a proton, it came from the energy contained within the space that was consumed. Scientists acknowledge that space contains energy. The Energy Density of space has already been documented in Equation 15 – Energy Density, in the chapter on Space Consumption. That equation is duplicated here without the 4π, and that makes it the Energy Density of the space consumed:

$$E_d = \frac{c^2}{G} \quad \frac{\text{speed of light squared}}{\text{space consumed}}$$

This equation may not clearly show why space contains the specified amount of energy, so let's make a minor adjustment for clarity. Then we will see that the energy density of space is equivalent to all the properties of space combined. Let's convert the c^2 to its dormant equivalent $1/e_o\mu_o$

$$\frac{1}{E_d} = \frac{1}{G\, e_o\, \mu_o}$$

$$E_d = 1.34661e^{27} \quad \text{Energy Density} \quad \frac{\text{Kg}}{\text{M}}$$

So, now when the space consumed by an electron in one second, is multiplied by the energy density of space, we can see the amount of energy released into an electromagnetic field in this transition. The calculations for an electron and proton are shown here:

$$\text{Energy Released} = \text{space consumed} \quad * \quad E_d \qquad \frac{Kg\,m^2}{s^2}$$

$$\text{Electron Energy} = \quad 6.07978e^{-41} \quad * \quad 1.34661e^{27} \; = \; 8.1871e^{-14}$$

$$\text{Proton Energy} = \quad 1.11634e^{-37} \quad * \quad 1.34661e^{27} \; = \; 1.5032e^{-10}$$

The energy released by both of these particles is exactly the same as the energy of the particles themselves. Note that the energy released does not come from the fundamental particle itself, but from the space that was consumed by the particle. If that energy came from the particle itself, it would immediately collapse. The particle only acts as a catalyst for the Space Transformation.

Energy Density of Charge

Once the energy is released from the consumed space, we can calculate the Energy Density of the outgoing Charges. We can calculate this by dividing the Energy of the Charge by the volume of space affected by the Charge. The volume of space affected by each Charge can be calculated by multiplying each Charge by the Gravitic Constant (G). First lets calculate all the volumes of space affected by their associated Charges.

		Gravitic Constant		Electron Charge		Charge Volume	
Gravitic	=	$6.6742e^{-11}$	*	$9.10938e^{-31}$	=	$6.07978e^{-41}$	m^3/s^2
Magnetic	=	$6.6742e^{-11}$	*	$2.81794e^{-8}$	=	$1.88075e^{-18}$	m^3/s^4
Electric	=	$6.6742e^{-11}$	*	$3.79467e^{12}$	=	253.364	m^3/s^2

The volume of space associated with Gravitic Charge is the volume of space consumed. The volume of space associated with the Magnetic Charge is the Magnetic Flux surrounding the fundamental particle. In this case we are calculating for an electron. The volume of space associated with the Electric Charge is the space containing the Electromagnetic Flux that was generated by the fundamental particle. Remember the difference in Charge volumes, Electric / Gravitic ($253.264 / 6.07978e^{-41}$) is equal to $4.16567e^{42}$. Now we can calculate the amount of energy per volume of space for each of these Charges. We referred to this as the Energy Density of Charge.

		Electron Energy		Charge Volume		Energy Density	
Gravitic	$=$	$1.96537e^{-56}$	/	$6.07978e^{-41}$	$=$	$3.23264e^{-16}$	Kg/m
Magnetic	$=$	$6.07978e^{-34}$	/	$1.88075e^{-18}$	$=$	$3.23264e^{-16}$	Kg/m
Electric	$=$	$8.18710e^{-14}$	/	253.364	$=$	$3.23264e^{-16}$	Kg/m

Here we see a most interesting phenomenon. The Energy Densities of the Charges are all the same. Apparently the energy is distributed evenly for the Gravitic, Magnetic, and Electric Fields. This means the Energy Density of a Magnetic Field is the same as its Electric Field. The difference is in the size of the field.

The Energy Density of the consumed space was $1.34661e^{27}$, whereas the Energy Density of the Electromagnetic Charge is $3.23264e^{-16}$. The difference between these two densities is $4.16567e^{42}$, as might be expected. This is because the volume of the Electromagnetic Field is $4.16567e^{42}$ times greater than the volume of space consumed. So it would appear that when all the dormant energy contained within the consumed space is activated, it is spread out over a much larger area.

How Magnetic Charge Works

We can refer to Figure 57 to help us visualize the transformation as it takes place. So, let's now go through this process one step at a time.

The Bohr Radius is the physical radius of an electron. The classical electron radius is the radius of an electromagnetic wave. The fine-structure constant squared is the difference between the Bohr Radius and the classical electron radius. It's the consumed space that is being transformed into energy, and that takes place just between the Bohr radius and Classical

Electron Radius. This space could be consumed at the Compton Radius which is between the Bohr and Classical Radii.

Envision a swirling magnetic fluxation of space, a small vortex within the heart of an electron, sucking up space, and funneling it through a portal as tiny as the classical electron radius. Out of the other end comes an electromagnetic wave. This is the transformation that continually takes place, second after second, day after day, on and on. This is the transformation of the pent-up energy contained within space, into an electromagnetic wave. We call it Electric and Magnetic Charge. These Charges make all chemical interactions possible.

Figure 57 – Bohr and Electron Radii

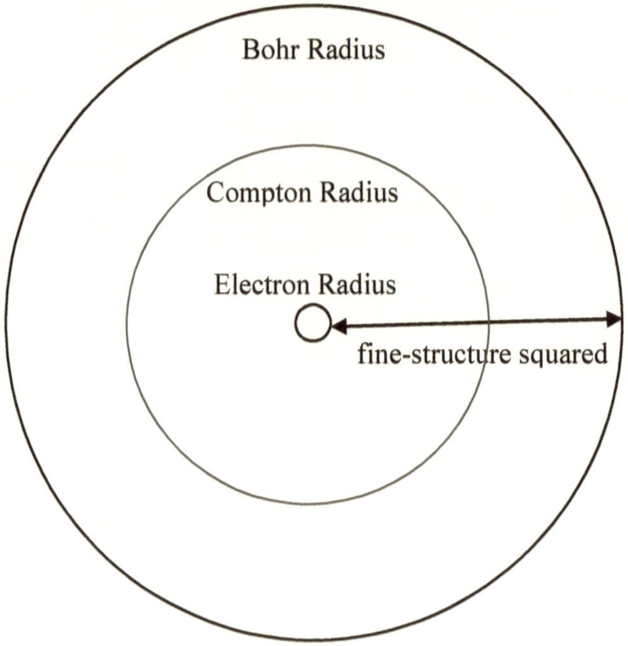

Once a fundamental particle of matter has been made, it consumes space. The amount of space consumed is what gives the particle the property of mass. As space is consumed, the permeability contained in that space is transformed into a magnetic flux. The charge of the magnetic flux is given by the expression (R_x / $1e^{-7}$). When the charge is multiplied by the Gravitic Constant, the result is the volume of the magnetic flux.

When matter is created, because of the rotation of the electron wave or proton wave, a magnetic flux is induced within the particles Bohr radius. Naturally, that magnetic flux utilizes the latent energy of the permeability properties contained within the space within which it (the magnetic flux) resides. Of course, it uses that energy to make its own wave.

Now that the magnetic flux is active, it sucks the latent energy out of the permittivity of the space it's in, and funnels it through the particles classical radius to make an electric wave. So, in this process, the properties of permeability and permittivity are sucked out of space.

One of the requirements of space is that it maintains its Energy Density at a constant level. The Energy Density of space is defined by its properties of space: permittivity, permeability, and gravity. (see the section - The Energy of Space). If space loses its properties, and becomes totally empty, it essentially has lost its essence, and it will collapse. It will simply collapse out of existence.

This explains a lot of things, and this is why magnetism has a direct connection to electricity and gravity. Also, magnetism is the strongest of forces as we shall see in the penultimate chapter. So it would appear that the driving "force" that causes the Gravitic and Electromagnetic Charges is the magnetic effects of the fundamental particle. It is the magnetic fluxation of space, that surrounds the fundamental particles, that drains the properties from space causing it to collapse. This is what causes gravity and creates the electromagnetic charge. It's all explained in this equation:

$$E_x = \frac{R_x \, r^2}{4\pi e_o \; 1e^{-7} \; G} \qquad \frac{Kg \, m^2}{s^2}$$

Chapter 34

The Fluxations of Space

Everything seems to be in flux.

The Anatomy of Space

The following equation is a copy of Equation 59 – Electromagnetic Energy, and is used now to show how we can calculate the electromagnetic force and also show the origins of the energy for this force. It is repeated here to show the simplicity of the converging properties of space. Keep this in mind as we go through the process of reviewing the transformations that take place.

$$E_x = \frac{R_x \quad r^2}{4\pi e_o \quad 1e^{-7} \quad G} \quad \frac{Kg \, m^2}{s^2}$$

In this equation we have a measurement for a volume of space in the numerator. In the denominator we have the three constants that describe the properties of space. The volume of space divided by the properties of that space, will reveal the characteristics of that space. What we have are three different fluxations of space. We are working with the properties of space and the fluxations of space to understand the characteristics of space. Note that no mass is specified anywhere in this equation. But, mass is the result of a fluxation of space.

The following equations are exactly the same, only with ovals encircling specific expressions. In the first equation two volumes of space are encircled. V_c is the volume of space consumed. V_e is the volume of space carrying the resulting Electromagnetic Charge. Notice that both volumes of space share the same portal R_x; the classical radius of a fundamental particle. The space that enters is consumed, as the electromagnetic charge exits.

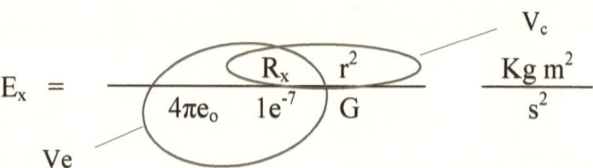

$$E_x = \frac{R_x \quad r^2}{4\pi e_o \quad 1e^{-7} \quad G} \quad \frac{Kg\ m^2}{s^2}$$

The following equation has three expressions encircled; these are for the three Charges of space: Q_1 – Gravitic (Mass), Q_2 – Magnetic, and Q_3 – Electromagnetic. Notice here also, that all three charges share the same portal R_x. Also, notice that the Magnetic Charge – Q_2 is embedded within the Electromagnetic Charge – Q_3.

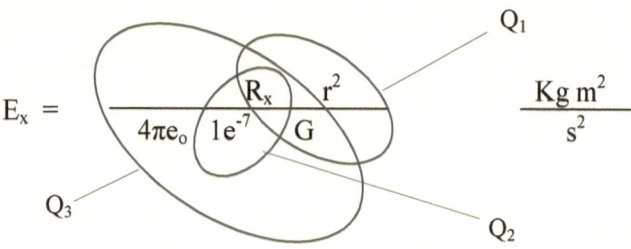

$$E_x = \frac{R_x \quad r^2}{4\pi e_o \quad 1e^{-7} \quad G} \quad \frac{Kg\ m^2}{s^2}$$

The following equation has only one expression encircled. This expression shows how the volume of space being consumed is converted into Magnetic Energy.

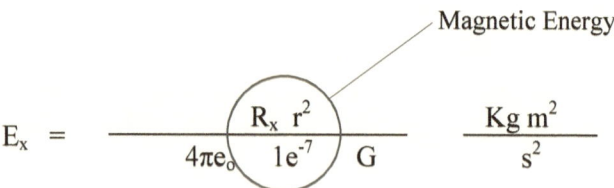

$$E_x = \frac{R_x \quad r^2}{4\pi e_o \quad 1e^{-7} \quad G} \quad \frac{Kg\ m^2}{s^2}$$

The next equation has two expression encircled. All of the properties in the denominator are the inverse of the energy density of the space consumed – V_c, in the numerator. This shows that the consumed space is highly packed with potential energy. The other encircled expression is also an Energy Density – E_d. But, this Energy Density pertains to the volume of space – V_e, which contains the resulting Electromagnetic Charge. So, the highly condensed energy in V_c is spread out throughout V_e.

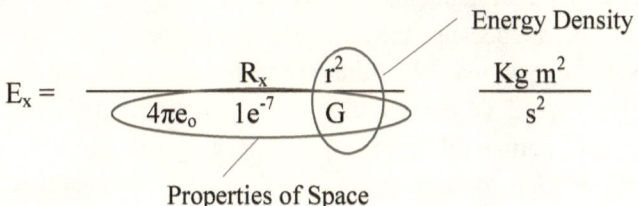

$$E_x = \frac{R_x}{4\pi e_o} \quad \frac{r^2}{1e^{-7}} \quad \frac{r^2}{G} \qquad \frac{Kg\ m^2}{s^2}$$

Energy Density

Properties of Space

The following Equation is exactly the same, but this time it is valued for an electron so calculations can be made.

$$E_x = \frac{2.81794e^{-15} * (1.46885e^{-13})^2}{4\pi\ 8.8542e^{-12} * 1e^{-7} * 6.6742e^{-11}} \qquad \frac{Kg\ m^2}{s^2}$$

$$E_x = \frac{6.07978e^{-41}}{7.42605e^{-28}} \qquad \frac{Kg\ m^2}{s^2}$$

$$E_x = 8.18710e^{-14} \qquad \frac{Kg\ m^2}{s^2}$$

The Dynamics of Space

All of these fluxations are occurring at the same time. Let's analyze what's happening in each case. We begin with the gravitational charge because that may to be the catalyst that initiates the other fluxations. For gravitational charge we have a volume of space in the numerator, divided by the Gravitic Constant. The result of this expression gives us the value of gravitic charge for this space in kilograms. The gravitic charge defines one of the characteristics of space surrounding a particle, and is referred to as mass. The volume of space in the numerator is totally consumed.

For gravitic charge the rate of change of volume of space tells us exactly what the quantity of mass is. In fact, this equation defines the capability of mass to attract other objects of mass. We could conclude that mass is the result of this gravitational fluxation of space. In a previous chapter we learned that mass has no real substance. This is exactly what we are seeing here, mass is a fluxation of space – the consuming of space. This pertains to all fundamental forms of mass: electrons, protons, and neutrons. This is true because they all consume space.

As space is consumed, two other space fluxations are kicked off. The first to be discussed is the creation of magnetic charge. As space is consumed, the permeability of that space goes through a Space Transformation. The dormant property of space - permeability ($1e^{-7}$), is activated into a dynamic magnetic field encircling the fundamental particle. This transformation is what gives magnetic charge to the particle. This is an ongoing and continuous process. As long as mass consumes space, the property of that space - permeability, will be transformed into a magnetic charge encircling the particle.

All fundamental particles go through this same process of creating a magnetic flux. The magnetic flux occupies the space surrounding the fundamental particles. The magnetic flux of fundamental particles overlap but will not oppose each other. Instead, they will position themselves to support each other. This keeps the electron and proton bound to each other in a stable position.

The second fluxation of space to be initiated is for the creation of electric charge. This charge results from the activity of the magnetic flux. As space is consumed, the permittivity of space goes through a conversion process. The magnetic flux draws out the permittivity of space and funnels it through the particles classical radius. The inactive property of permittivity is thus converted into a dynamic electric field. The radius of the electric field is the classical radius of the fundamental particle. Which means that electrons generate negative electric waves, whereas protons generate positive electric waves. That's why the electric charges of protons are positive and the electron charges of electrons are negative.

Both protons and electrons go through the same process of creating electric fields from the properties of space, one creates a positive electric field while the other creates a negative electric field. The fields are not only different in charge, but in radius size, which is considerable. In normal conditions, when an electron and proton are in close proximity, these electric fields of opposite charge will attract each other and combine into a normal neutral electromagnetic field. Electric fields of the same charge will oppose each other making them resisting charges.

The creation of electric charges is also an ongoing and continuous process. As long as mass consumes space, the property of permittivity, will be transformed into an electric field. What we are seeing in all three cases are the fluxations of space due to the different properties of space. All interactions that take place between particles of matter are the results of these fluxations of space. The electric and magnetic charge equations show a specific rate of change in the charge of space. The gravitational charge equation gives a specific rate of space consumption.

We also need to remember that the structure of electric and magnetic fields are different. The electric charge of a particle can be positive or negative. The structure of the electric field can be pictured as coming in straight toward the particle, or going directly out from the particle. Magnetic charge is polar so the charge has a north and a south pole. The structure of magnetic fields follow a curved path from the particle that circles back into itself. Therefore the magnetic field does not have the same "reach" as the electric field.

Matter in Motion

These fluxations of space govern the motion of all bodies in space. Gravitational flux is the flow of space toward all objects of mass. The expansion of space outward from all objects of mass will also impart the power of acceleration to all bodies in the universe.

Electric flux is the flow of an electric field and magnetic flux is the flow of a magnetic field. These fluxations enable electromagnetic fields to surround the fundamental particles of matter creating electric and magnetic charges. These two fluxations govern the motion of all particles of matter.

What we're seeing at the quantum level of the fundamental particles are the effects of the dynamics of space. At the radius of the fundamental particles we see space being consumed. The effect of this consumption is gravity. Space consumption accelerates all objects toward the source of consumption.

It's the permeability and permittivity of space that permits electromagnetic waves to flow through the particles creating electric and magnetic charges. The electromagnetic fields of engaging particles cause them to accelerate. The fields of two particles with the same charge will repel each other. The fields of two particles with opposite charges will attract each other. The forces of nature are the result of the fluxations of space.

Electric and magnetic charges are static in the sense that their values are constant. These charges are continually renewed as space expands and contracts. The flux of space is generated by the dynamics of space (the contraction and expansion) surrounding fundamental particles. The energy of these charges does not come from the particle itself, but from the energy within the space surrounding the particle.

Einstein told us that the structure of space determines the motion of matter. That statement is true, and not only regarding gravity; it is applicable to the electromagnetic forces as well. This explains why magnets do not quickly lose their strength after repeated uses. The source of a magnet's strength is continually renewed. As space continually contracts, the magnetic field and electric field continues flowing. As space continually contracts, the flowing of the electric and magnetic charge is maintained.

This is analogous to celestial bodies. The charges of space contraction and space expansion are unrelenting, causing planets and galaxies to revolve and rotate. But it's all done through laws of motion, without the energy supply being consumed or depleted. So it is in the subatomic world. The fundamental particles of matter are in constant motion as they interact with each other and move through space according to the properties of that space. Everything is in constant motion, moved by an inexhaustible supply of energy perpetually renewed as space forever expands and contracts.

Converging Forces

Figure 58 – Converging Forces, shows how these three properties of space converge with each other. Think of these properties of space being at right angles to each other. The Gravitational Constant (G) represents Gravity. The Permeability of Space Constant (e^7) represents Magnetism. Coulomb's Constant (K_c) represents Electricity. The Gravitational Constant (G) multiplied by the Permeability of Space Constant (e^7) is equal to $6.6742e^{-4}$ m^2/s^4 – (G_m). This value is acceleration squared; it is the combination of two properties of space – Gravity and Permeability. This is the rate of acceleration of the magnetic field surrounding a fundamental particle. We shall see the significance of this in a later chapter. Coulomb's Constant (K_c) multiplied by the Permeability of Space Constant (e^7) is equal to c^2; which is the speed of light squared. This is the speed at which an electromagnetic field propagates out from a fundamental particle.

Figure 58 – Converging Forces

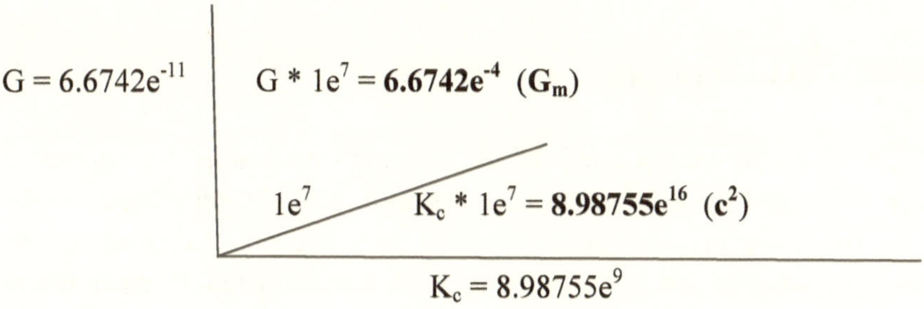

$G = 6.6742e^{-11}$ $G * 1e^7 = 6.6742e^{-4}$ (G_m)

$1e^7$ $K_c * 1e^7 = 8.98755e^{16}$ (c^2)

$K_c = 8.98755e^9$

Table 36 – Charges, shows how these numbers fit into the equations for the Charges of Force. All Charges can be defined by a volume of space divided by the Gravitational Constant. The volume of space for Q_1 is the volume of space consumed. The volume of space for Q_2 is the volume of the magnetic field. The volume of space for Q_3 is the volume of the electromagnetic field.

Table 36 – Charges

$$Q_1 = \frac{R_e\, r^2}{G} = \frac{2.8179e^{-15} * 2.15753e^{-26}}{6.6742e^{-11}} = 9.10938e^{-31}\ K_g$$

$$Q_2 = \frac{R_e\, G_m}{G} = \frac{2.8179e^{-15} * 6.6742e^{-4}}{6.6742e^{-11}} = 2.81794e^{-8}\ \frac{K_g}{s^2}$$

$$Q_3 = \frac{R_e\, c^2}{G} = \frac{2.8179e^{-15} * 8.9875e^{16}}{6.6742e^{-11}} = 3.79467e^{+12}\ K_g$$

The one common thread that all three Charges have in their volume of space, is the classical radius of a fundamental particle. In this case we are using the classical radius of an electron in our examples. The origin of these Charges appears to be at the radius of the fundamental particles. All three Charges are defined in kilograms. So, we can compare their relative strength by kilograms of mass. The strength of the Magnetic Flux is $3.09345e^{22}$ times greater than the mass of an electron ($9.10938e^{-31}$ Kilograms). The strength of the Electric Flux is $4.16567e^{42}$ times greater than the mass of an electron.

Gravitic Constant Redefined

Up until now we have defined the Gravitic Constant as the amount of space consumed by a kilogram of mass. But, a slight adjustment to the equations in the previous section shows there are other alternatives. In Table 37, we can see that the volume of space related to each force, divided by the Charge of that force is equal to the Gravitic Constant. This means that the Gravitic Constant can be defined by all three forces.

Table 37 – Gravitic Constant

$$G = \frac{R_e \, r^2}{Q_1} = \frac{2.8179e^{-15} * 2.15753e^{-26}}{9.10938e^{-31}} = 6.6742e^{-11} \; \frac{m^3}{K_g \, s^2}$$

$$G = \frac{R_e \, G_m}{Q_2} = \frac{2.8179e^{-15} * 6.6742e^{-4}}{2.81794e^{-8}} = 6.6742e^{-11} \; \frac{m^3}{K_g \, s^2}$$

$$G = \frac{R_e \, c^2}{Q_3} = \frac{2.8179e^{-15} * 8.9875e^{16}}{3.79467e^{+12}} = 6.6742e^{-11} \; \frac{m^3}{K_g \, s^2}$$

The Gravitic Constant can be defined as it was in the chapter on Space Contraction; the volume of space consumed per charge of mass. The second alternative is to define the Gravitic Constant as the volume of space containing the Magnetic Flux generated by the magnetic charge. The third alternative definition for the Gravitic Constant, is the amount of space containing the Electromagnetic Flux generated by the electromagnetic charge.

Space Fluxations

We saw in a previous chapter, (The Unification of Forces), that we could calculate the Electromagnetic Force by multiplying the Gravitic Constant, the Gravitic Charge, and the Electric Charge, as shown in the following equation: Force = $G * Q_1 * Q_3$. Let's substitute the expressions from Table 36 – Charges, for Q_1 and Q_3 into a new equation:

$$\text{Force} = G * \frac{R_e r^2}{G} * \frac{R_e c^2}{G}$$

Both numerators of these expressions are measurements of volumes of space. The first($R_e r^2$) is the volume of space consumed by an electron. The second ($R_e c^2$) is the volume of space holding the generated electromagnetic field. So let's revise this equation to show the volumes and factor out the Gravitic Constant (G) from the numerator and denominator. So now we have a new equation (Equation 60) that shows how the Electromagnetic Force is generated.

Equation 60 – Electromagnetic Force II

$$\text{Force} = \frac{V_c * V_e}{G}$$

Where:

V_c is the volume of space consumed.
V_e is the volume of electromagnetic space.
G is the Gravitic Constant

But first, before an explanation, let's validate this equation by using it to calculate the Electromagnetic Force of an electron. V_c is the volume of space consumed and V_e is the volume of space containing the generated electromagnetic field.

$$\text{Force} = \frac{6.07978e^{-41} * 253.264}{6.6742e^{-11}} = 2.30708e^{-28} \quad \frac{K_g\, m^3}{s^2}$$

Yes, it works, we have the correct value for the Electromagnetic Force. Now we can interpret this equation to see how the Electromagnetic Force is generated. All the dormant energy from within the consumed space (V_c) is transformed into electromagnetic energy, and spread throughout the surrounding field (V_e). This is all done as specified by the Gravitic Constant (G). The space (V_c) is consumed by the mass as specified by the Gravitic Constant. The resulting electromagnetic field occupies the space (V_e) as specified by the third definition of the Gravitic Constant. The electromagnetic field (V_e) is $4.16567e^{42}$ times greater than the space consumed (V_c).

In the previous chapter we had calculated the latent energy contained in the space consumed. That calculation is repeated here for convenience.

$$\frac{\text{space consumed}}{G\, e_o\, \mu_o} \qquad \frac{m^3 / s^2}{m / Kg}$$

$$\frac{6.07978e^{-41}}{7.42605e^{-28}} \quad = \quad 8.1871e^{-14} \qquad \frac{Kg\, m^2}{s^2}$$

The latent energy in that space goes through what we called a Space Transformation to convert it into active energy. Energy is not created in this process, just transformed, and spread throughout a larger area. This energy can be calculated in two additional ways. Energy is equal to Charge times velocity squared. The energy consumed by an electron is equal to the Charge (mass) of an electron multiplied by the velocity at which the energy is dispersed squared. Remember that gravity and electromagnetism obey the Law of Inverse Squares. That's why the velocity is squared. The energy dispersed by an electron is equal to the Electromagnetic Charge times the velocity at which the energy is activated squared. These calculations are shown below:

$$\frac{\text{Energy}}{8.1871e^{-14}} \quad = \quad \frac{\text{Charge}}{9.10938e^{-31}} \quad * \quad \frac{(\text{velocity})^2}{(299{,}792{,}458)^2}$$
$$8.1871e^{-14} \quad = \quad 3.79467e^{12} \quad * \quad (1.46885e^{-13})^2$$

The electromagnetic charge is $4.16567e^{42}$ times greater than the gravitic charge, and the velocity squared of the energy dispersion is

$4.16567e^{42}$ times greater than velocity squared of space consumption. So, there appears to be a perfect balance in nature and in this transformation.

Notice that in Equation 60, that there is no mass. Mass (the Gravitic Charge), the Magnetic Charge, and the Electric Charge are all created by space fluxations. Everything that exists was created from nothing. But, will we ever run out of space? No! Space is being created much faster than it can be consumed.

Creating Charge

Let's take a few moments and take a closer look at how all the Charges are created by these fluxations. We will show this process step by step. This does not mean that the process is a sequence of steps. Once the process is started, every step of the process is occurring at the same time.

All of the fluxations of space occur at a specific location – the classic radius of a fundamental particle. Where no fundamental particles exists in space, nothing happens. Where there is mass, everything happens, and it all happens at the classical radius of the fundamental particles.

The first question we can ask ourselves is, "Where did the mass come from"? The chapter entitled, "The Creation of Matter" shows that electrons and protons were created from the energy of empty space. We can know this from a very simple equation, space divided by the Gravitic Constant is equal to the Charge of Mass. The following two calculations show the creation of an electron and proton:

Particle:	Volume	÷	G	=	Charge (Q_1)	
Electron:	$6.07978e^{-41}$	÷	$6.6742e^{-11}$	=	$9.10938e^{-31}$	K_g
Proton:	$1.11634e^{-37}$	÷	$6.6742e^{-11}$	=	$1.67262e^{-27}$	K_g

After mass has been created the following three steps are the three processes that create the Electric and Magnetic Charges. These three processes are the work of the three properties of space. At the classical radius of each of these particles, space is consumed, and a Magnetic Flux is created from/by the Permeability of that space. This Magnetic Flux is called Magnetic Charge (Q_2). See the following calculation:

$$Q_2 = \frac{R_e}{1e^{-7}} = \frac{2.81794e^{-15}}{1e^{-7}} = 2.81794e^{-8} \; \frac{K_g}{s^2}$$

That Magnetic Flux is processed by the Permittivity of that consumed space to produce a much larger Electromagnetic Flux. Remember, the Magnetic Flux is embedded within the Electromagnetic Flux. The following calculation shows the expansion of the Electromagnetic Flux.

$$V_e = \frac{Q_2}{4\pi e} = \frac{2.81794e^{-8}}{1.11265e^{-10}} = 253.264 \; \frac{m^3}{s^2}$$

Finally, we can divide that electromagnetic space by the Gravitic Constant to get the Electric Charge (Q_3) in terms of Kilograms of mass.

$$Q_3 = \frac{V_e}{G} = \frac{253.264}{6.6742e^{-11}} = 3.79467e^{+12} \; K_g$$

All three of these steps are automatically included in Equation 60 – Electromagnetic Force II. The numerator of Equation 60, contains just two volumes of space; V_c is the space consumed, and V_e is the volume of space containing the Electromagnetic field. Think of both volumes of space sharing the same classical radius. A volume of space (V_c) is being funneled into the radius and consumed out of existence. Out from the other side of the radius are the generated electric and magnetic fields.

Figure 59 – Space Fluxations

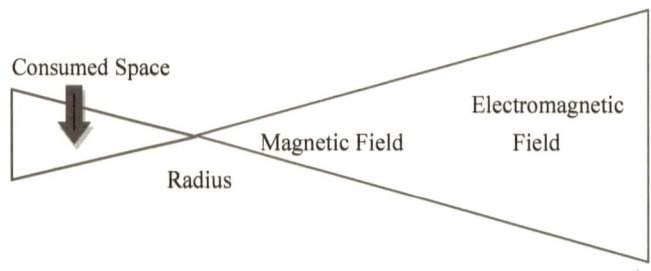

The Origin of Everything

The following is a short synopsis of how everything came into being.
A long, long time ago, before the stars, before the galaxies, before any planets, even before time and space existed, there was no universe. Then God said, "let there be light". Then the Word of God made space, and put light in the space that was just created. And the light shined and that was the first cosmic day.

After observing the light, then God said, let's separate the light from the darkness and let the light flow. So, the Word of God gave the space the properties to let the light flow through it, and he called it darkness, or dark energy as we know it. The light flowed freely in the space, and that was the second cosmic day.

After gazing upon the flowing light, God said. "This is good!, now let's stretch out the heavens and fill the universe with light." Then the Word of God stretched out the heavens, expanding the universe, and the light flowed, filling the universe. This was the third cosmic day.

After observing the universe filled with light, God said, "This is good!, Now let's fill the universe with stuff." So the Word of God, set into motion some laws of nature, where mass would be continuously created at the edge of the universe. So, as the universe would continue to expand, so mass would be continually added. God obviously didn't want to run out of stuff. And this was the fourth cosmic day.

Then God saw all the stuff that was made and said, "This is good stuff! Now let all this stuff come together to make stars and galaxies, and really make the heavens sparkle." So, the Word of God caused the stuff to consume the dark space, and release its stored energy. This caused the stuff to come together and make stars and galaxies to light up the universe. And many of the galaxies were made to have beautiful spiral arms to keep track of time, like giant clocks. This was the fifth cosmic day.

And then God lifted His eyes to behold the heavens and all the shining stars within the galaxies, and He was very pleased. Then He said, "This is really good stuff!" Now let there be planets and moons and other kinds of things that travel in space to light up the night sky. Because we're going to put people on one of these planets, and it needs to be prepared to bear life. So, the Word of God set some laws in motion so that as the mass consumed the darkness, it would also convert some of the darkness into energy. So the Word made different kinds of energy, one with great strength at small distances so that many different kinds of mass could be created, and another so that all kinds of life could be created. And these different kinds of energy would work together so that many kinds of life could be brought into existence. And that was the sixth cosmic day.

And when God lifted up His eyes on the seventh cosmic day, and beheld what was made, His heart was filled with joy. And He said, "This is very, very good, I am well pleased. We shall rest now on this seventh cosmic day to enjoy all the work that has been done." And then all the angels shouted for joy, with great praise and shouts of acclamation. Tens of thousands of angels sang in harmonious praises to God for joy of what had been done; knowing that God had great plans for His creation. If we only knew.

Matter in Random Motion

The following is a short scientific explanation of how everything came into being. This is for scientists who don't believe God exists or was involved in any way in the creation of the universe.

The universe appears to have begun with expanding space and dark energy. It would seem that, as space expands, the dark energy within remains the same, never exhausted nor diluted. Due to the expansion of space, the dark energy that permeates the universe becomes matter at the edge of the universe. Matter consumes space even as space expands outward from that same matter. These two fluxations induce acceleration. They also enable the electric and magnetic fluxations of space, which in turn enable electromagnetic fields to surround all particles of matter, thus creating electrically and magnetically charged particles.

When this universe was created, there was no excitement, there were no shouts of joy, because there was no one there to observe it. It just happened – all by itself. The life that spontaneously originated from the muck of swamps was no big deal. Just aimless life, with nothing to live for, and no reason to be. But, the life got more complex and the complex life devoured the simple life. It all had no point or purpose so, there is no joy or happiness, because there is no point to anything, it's all just a matter in random motion.

Chapter 35

The Strong Nuclear Force

There is no force like the nuclear force.

The Mysterious Nuclear Force

One of the most mysterious, and least understood, forces in the universe is the strong nuclear force. This force, along with its close relative the weak nuclear force, are relatively new on the scientific scene. It was only after Bohr's new model of the atom (see chapter 27 – The Nature of Matter) became firmly established in the 20th century that it became necessary to explain how the nucleus of an atom remains intact when the powerful electric charges of its protons are forcing them apart. With no force to bind the protons together, the atom would literally fly apart.

Before examining the nature of the strong nuclear force let's compare the relatives sizes of a proton and an electron to get a better perspective of proportion. The radius of an electron is 1836 times larger than that of a proton, so there is plenty of room for the proton to wander around within the midst of an electron shell. This would be comparable in size to a baseball in the middle of a major league baseball park. The most massive atoms found in nature are of uranium – U238. But, there are seven electron shells to the uranium atom. This would be comparable in size to a bag of 238 small marbles in the middle of a major league baseball park. There is plenty of room for the protons within the electron shells, however elements with larger nuclei than uranium are unstable and disintegrate rapidly into smaller elements.

In a previous chapter we calculated the acceleration of a proton one meter away from another charged particle. This acceleration is .13793 m/s^2, about 13.79 centimeters or 5½ inches per-second squared, an astonishingly powerful force and a phenomenal acceleration considering the infinitesimal size of a single proton. What then keeps the protons in the nucleus of an atom together when the natural tendency is for the protons to explode outward from one another?

Unable to answer this question through conventional means, the existence of a previously unknown force was theorized, something stronger than the repelling forces of electric charge. This theoretical force became known as the strong nuclear force. This force makes no appearance outside the nucleus of an atom, and has no affect on the surrounding electrons. This force, although exceedingly powerful, has a range only slightly larger than the diameter of the proton itself.

Other than the necessity for its existence, there is virtually no understanding of the strong nuclear force. There are few theoretical explanations for the origin of this force, none of them widely accepted by the scientific community. All that can be said is that there exists an extremely powerful force acting only on protons and neutrons, and only for a short distance. What is the source of this powerful force?

Perhaps the most prominent of the nuclear force theories was presented in 1935 by Japanese physicist Hideki Yukawa (1907-1981). Yukawa proposed that unknown particles, with a mass 200 times that of an electron, were being absorbed and ejected from the neutrons and protons in the nucleus of atoms. The life span of these particles was expected to be very brief, only one billionth of a second. Yukawa calculated that the exchange of those particles would produce an attraction sufficient to hold the nucleus together. Although the theory seems bazaar, two particles fitting Yukawa's description: one called a muon, the other a pi meson, were discovered twelve years later at a high-altitude location using extremely sensitive photographic plates. The discovery of these particles lent creditability to Yukawa's theory, and in 1949 he was the first Japanese scientist to be awarded the Nobel prize.[57]

Although Yukawa's theory is fascinating it does not contain the answers we seek regarding the force that binds the nucleus of atoms. Particles 200 times more massive than an electron bouncing around in the nucleus of an atom would only cause greater instability. When the muons die, how are they replenished? What happens when there are too many muons in the nucleus, or not enough? There must always be at least one muon in every nucleus of every atom to keep it balanced. How is the balance of muons achieved and controlled? Where do muons come from? The detection of the muons themselves shows there are too few of them to accommodate all atoms, and there is no evidence whatsoever of them having a permanent residence within atoms. More questions are raised than answered with this theory. Also note that Yukawa's Nobel prize was awarded for predicting the existence of particles larger than the electron, not the explanation of the nuclear force.

Capacitating Charges

Before proceeding with the explanation of the nuclear force, let's temporarily turn our attention toward the electron. We have already calculated the acceleration of a proton one meter away from another charged particle (.13793 m/s^2, about 13.79 centimeters or 5½ inches per-second squared). The acceleration of an electron resulting from another charged particle one meter away is 253.26 m/s^2, about 277 yards per second squared, a very high acceleration. The same magnitude of force exists on the electron as the proton, but we must also consider how light and mobile the electron is. With a mass of only 9.1094x10^{-31} kilograms, an electron is about 1,836 times lighter than a proton; therefore its acceleration is also 1,836 times greater than that of a proton.

Considering the magnitude of force causing the acceleration of electrons, how is it that billions of them can be packed together into a small area? That is exactly what an electronic device called a capacitor is capable of. Capacitors are named such because of their capacity to hold electric charge. Typically made of thin layers of two materials, one layer is a metal like aluminum through which electric current can flow; the other is a dielectric material which does not conduct electricity. These layers are either alternately stacked on top of each other, or stacked and rolled into cylinders enabling them to fit easily on electronic circuit boards.

When an electric current runs through the conducting metal it attracts and holds electrons. This accumulation of electrons is the buildup of electric charge. Even after the current is terminated, the collected electric charge remains intact. These little devices (capacitors) can hold billions and billions of electrons in a relatively small confined area. How is this possible when powerful electric charges are "pushing" the electrons away from each other? Note that capacitors do explode when charged beyond their capacity.

What force can possibly override the electric charge of all those electrons? The answer to that question is magnetism. Remember, electrons also have a magnetic field. It's magnetic attraction that holds electrons to the surface of the metal. We shall soon see that this magnetic attraction is strong enough to override the repelling force of the electric charge.

Invoking Ockham's Razor

It's time to apply the principle of Ockham's razor to the strong nuclear force. If there exists an explanation of the phenomenon that binds the nuclei of atoms together via known forces of nature, then there is no need to conjure up a mysterious force.

Today's theories suggest that the strong nuclear force interacts solely between protons and neutrons over a very short distance. Conduct your own scientific experiment. Hold two equally sized magnets in your fingers, one magnet in each hand, with the poles facing the same direction. Slowly bring them together. When the magnets are within about one diameter distance, one magnet will suddenly "jump" out of your fingers and attach itself to the other magnet. Magnetism is a powerful force effective at a very short range. What we have called the nuclear force is nothing more than magnetism!

The magnetic attraction of protons and neutrons has been discounted in the electromagnetic effect. The reminder here is that it is both an electro and a magnetic effect. The magnetic effect has been ignored. All three fundamental particles: the electron, proton, and neutron have a magnetic field. Only the electrons and protons have electric fields; these electric fields are of opposite charge. When scientific measurements are made on electric charge, they are mostly concerned with the electric charge of electrons. Located on the outer part of an atom, electrons flow freely, and are easier to measure than the charge on protons. When calculations are made to validate Coulomb's force, stationary charges and fixed distances are used. Coulomb's law is confirmed and magnetic effects do not appear to apply.

However, in the world of atomic nuclei magnetic fields are in effect. Both the proton and neutron have a magnetic field surrounding them. Also, the proton has a positive electric charge, while the neutron has no electric charge. This makes for an interesting combination.

The Nuclear Force

Let's remember what we learned about the magnetic force just a few chapters ago. We were able to calculate the size of the magnetic field by multiplying the Magnetic Charge by the Gravitic Constant. The calculation for the volume for the magnetic field is shown below:

$$V = \underset{\substack{\text{Gravitic} \\ \text{Constant}}}{6.6742e^{-11} \; \frac{m^3}{Kg\, s^2}} \quad * \quad \underset{\substack{\text{Magnetic} \\ \text{Charge}}}{2.81794e^{-8} \; \frac{Kg}{s^2}} \quad = \quad 1.88075e^{-18} \; \frac{m^3}{s^4}$$

Now notice the units of measurements of this magnetic volume of space. It's not just the normal acceleration of a space fluxation with s^2 in the denominator. This fluxation is at a rate of s^4. This is the magnetic flux. This is why magnets will jump out of your fingers when attracted by another magnet. It's strength drops off sharply with distance, but rapidly increases near the source. These are the characteristics of the Strong Nuclear Force. Examine Figure 60 – Fundamental Forces, to see the relative strengths of these three forces.

Figure 60 – Fundamental Forces

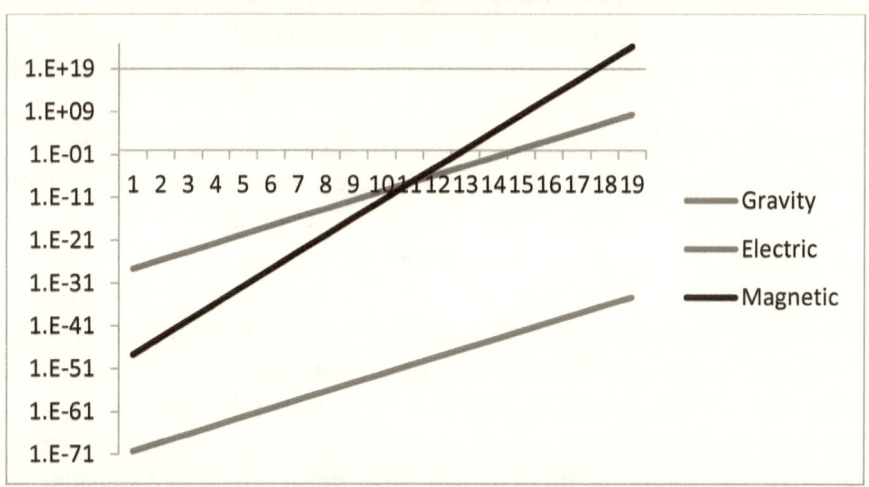

Notice at the left, one meter from the source, all three forces are at their calculated strength. As we move to the right, closer to the source, all three forces get stronger, however the magnetic force gets stronger faster. At mid-point the electric and magnetic forces are about equal. This is at a distance of $8.6175e\text{-}11e^{-11}$ meters. This is slightly larger than the Bohr Radius $(5.29177e^{-11})$ of an electron. So, it would appear that chemical interactions are controlled by both magnetic and electric forces; with the magnetic force being slightly more dominate at the Bohr Radius.

When the forces are near the Bohr radius of the Protons and Neutrons, $1e^{-14}$, at the nucleus of an atom, the magnetic force becomes very much stronger than the electric force. Thousands of times stronger! That's when the magnetic force becomes the Strong Nuclear Force. So, there is no mysterious Strong Nuclear Force. The so called Strong Nuclear Force is really just the Magnetic Force at close range. So now, the connection between the four known forces of the universe, have been clearly identified.

Summary

We need not discuss the weak nuclear force. The hypothesis being made here is that the weak nuclear force is simply the result of the interactions of the electromagnetic forces. All of these interactions may not be currently known or understood; more research is required. But Weinberg, Abdus Salam, and Glashow, have connected the weak nuclear force to the electromagnetic force. They were awarded the Nobel prize for making this connection, which is now known as the Electroweak Unification.

In this chapter we have presented a strong case that the nuclear force is really the magnetic force in disguise. We have eliminated the need for both the strong and the weak nuclear forces. That leaves us with only the fluxations of space as the source of the forces of nature.

The forces of nature have now been explain. However, there is still one force that remains a mystery; it is the Fundamental Force. This is the force that is causing the expansion of the universe and the creation of space itself. What caused the Big Flash that created the universe? This is the one Force that powers the whole universe. In order to create the entire universe, the source of this creative force must come from outside the universe. It must be a source of power from God!

Chapter 36

The Fundamental Force

There is one force that powers the universe.

The Structure of Space

This treatise on the workings of the physical universe is now complete. We have explored mysteries at both an atomic and a galactic scale. In some cases their solutions were thought to have been known, in other cases enlightenment is still fervently sought. Although some of these phenomena seem unrelated, they are bound together by one scientific principle: the expansion and contraction of space. Having seen a variety of these phenomena explained, our picture of the universe is now much clearer than when we began. It is the purpose of this chapter to summarize our new knowledge.

Beginning with the first chapter, 'On the Shoulders of Giants', we take a tour through the history of human understanding of gravity from Galileo to Einstein. The second chapter is a brief review of, 'Kepler's Laws', which play an important part in our understanding of how the universe works. This is one of the foundations which we build upon to increase our depth of understanding. The other foundations are Newton's laws and Einstein's relativity.

Space Contraction

Our new understanding of gravity varies considerably from the Newtonian view. Einstein showed us that gravity is not a force, but an effect, stating that the structure of space determines the motion of matter. We have proceeded a step farther, realizing that the structure of space is itself an effect dictated by its own contraction and expansion.

In the next six chapters, (3 - 8), we learned how Gravity works. We learn that matter consumes space, and as it does the surrounding space is drawn inward, creating a perpetual flow of space toward matter, the effect of which we call gravity. This makes clear how all matter attracts all other matter in the universe.

Space Expansion

The chapter on, 'Dark Energy', is the introduction to expanding space and the characteristics of dark energy. This chapter shows how we know the universe is expanding and brings us up-to-date with the current understanding of dark energy.

Chapter 10 explains the problem of the 'Pioneer Anomaly'. Two Pioneer spacecraft launched in the early 70's have long since traveled beyond the solar system. Their whereabouts have been tracked; strangely, they are not on their expected course according to Kepler, Newton, nor Einstein. An unaccounted for acceleration toward the sun has drawn them off course. We have identified the source of the excess acceleration as the result of space expansion. In the future, spacecraft can be tracked much more accurately by accounting for this additional assist in acceleration.

The stars at the perimeter of the galaxies rotate at a much greater velocity than predicted by Kepler's laws. This and the existence of spiral arms enduring billions of years have confounded astronomers. Physicists have been searching for the elusive dark matter to explain these phenomenon; there is no longer a need to go on searching. These apparent instances of disregard for the established laws of physics are now explainable, and discussed in the three chapters on Galaxy Rotation, (chapters 11, 12 and 13). As with the Pioneer spacecraft, the additional acceleration of stars at the outer edge of the galaxy, as well as the rotation of the galaxies themselves, is induced by the expansion of space.

Like the search for dark matter, the MOND theory, which speculates on the apparent increase of gravitational attraction over great distances, can now be explained. The expansion of space lends additional acceleration to remote bodies, causing gravitational attraction to appear stronger with distance.

Five chapters, (14 - 18) explain how the Tides work. We learn that gravity cannot be the cause of the tidal effect. The expansion and contraction of space dictate the structure of space, which, in turn, determine the shape of objects occupying that space. When two objects come into close proximity, both are slightly distorted, taking on an ellipsoidal shape. This is the tidal effect, observed on both sides of the Earth, and is a direct result of the principles of space expansion and contraction at work. This also explains why the tidal effect diminishes with the distance cubed.

Also, it's the ellipsoidal shape of satellites that keep them tidally locked to their parent planet. The tidal bulge of a satellite comes to stability when always facing the parent planet. It's the tidal effect that keeps the same side of the moon always facing the Earth.

The precession of planetary orbits is thought to have been an effect of gravitational attraction between planets that come into conjunction over the normal course of their orbits. In the two chapters, 'Planets in Precession' and 'Planetary Perturbations', (chapters 19 and 20), we learn that such minor displacements of the orbits (planetary perturbations) are hardly enough to cause a true precession. It is the slight assist in acceleration imparted by the expansion of space which produces the precessions observed in the planetary orbits.

The nearly circular orbits of planets and moons is another aspect of the universe illuminated by the principle of space expansion. The chapter on, 'Circular Orbits', (chapter 21) reveals how space expansion gives a slight boost in acceleration to objects in orbit at their aphelion and a slight deceleration at the perihelion. This slight adjustment will shorten the aphelion and lengthen the perihelion. This means that with each revolution of a planet around its parent star, its orbit becomes slightly more circular. This principle also applies to moons in orbit around planets.

Though the phenomena described above may appear to be unrelated, all are impacted by the same scientific principle – the contraction and expansion of space. All motion of bodies in space is determined by the same scientific principle – the structure of space.

The Expanding Universe

The three chapters on the Expanding Universe, (chapters 22-24) describe the expected fate of the universe as currently predicted by scientists depending on whether the universe is expanding, contracting, or static. Hubble's law is explained along with Hubble's Constant and the expansion of the universe. This law is explained in the standard way; however an unexpected twist was introduced. We saw that although the galaxies are moving outward toward the edge of the universe, but they are also accelerating inward toward the center of the universe. This new insight gives us a totally different view on the origin and ultimate fate of the universe.

Building the Universe

The chapters, 'The Expanding Universe', and 'The Making of Galaxies', (chapters 25-26), explains how the universe is expanding and shows us how galaxies are made as a result of that expansion. At the top of the universe, at the speed of light, energy attempting to push beyond the bounds of the universe is transformed into the fundamental building blocks of matter: – protons and electrons. All elements are formed from these particles, starting with hydrogen clouds falling from the perimeter of the universe and on to the heaviest elements born in the furnaces of the stars and scattered across the galaxy at their death.

Some of the most interesting experiments in science are described in the chapter, 'The Nature of Matter', (chapter 27), to show the true nature of matter. Once understood, the origin of matter is spelled out in detail in, 'The Creation of Matter', (chapter 28). We have seen how matter is continuously being created from the energy in space. Matter is created at the top of the universe and falls downward toward the center. We now possess a greater understanding of how galaxies are formed and the powers that rotate them. The expansion and contraction of space work together to guide the rotation not of just galaxies, but of the stars, planets and moons within them.

There are billions of galaxies containing billions of stars and planets; all in perpetual motion governed by the dynamics of space. All this motion is initiated without any force being applied to the objects. The universe is truly a marvelous and wonderful creation.

The Forces of Nature

The following seven chapters show how all the forces are unified. In the chapter on, 'The Properties of Space', (chapter 29), the two fundamental properties of space: Permeability and Permittivity are discussed. These two properties are keys to understanding the nature of Charge. Chapter 30, 'The Electromagnetic Force' is explained along with Coulombs Constant and Coulomb's Law. In chapter 31, 'The Nature of Charge', the three Charges are explained.

Once we understand the three Charges, then Chapter 32, 'The Unification of Forces', describes the forces of nature and how they are unified. Space Transformations, (chapter 33), is the chapter that reveals the transformations that take place from the potential energy of empty space to the dynamic energy of the Charges. The Fluxations of Space, (chapter 34) discusses the dynamics of space and the origin of everything.

Finally, in the penultimate chapter, The Strong Nuclear Force, (chapter 35), explains where the strong nuclear comes from. It seems to have been hidden away, but now its source becomes obvious. The source of the strong nuclear force is magnetism and it was there all the time, but we just didn't recognize it.

The Fundamental Force

All motion of matter is governed by the fluxations of space. These are the "forces" that drive all action within the universe. For years we have sought after the ultimate theory to unify the forces of nature. We have shown here that the fluxations of space come together in the heart of the fundamental particles to create the electromagnetic forces.

Of the ultimate unified theory, John Archibald Wheeler says: "It must be an utterly simple idea that when we discover it, it will be so compelling, so inevitable, and so beautiful; that we will all say to each other, Oh, how could it have been otherwise!"[58] And it is both simple and beautiful! I would repeat those same words regarding the design of the universe and the structure of space; "It's an utterly simple idea, so compelling, so inevitable, and so beautiful. Oh, how could it have been otherwise?"

The Grand Design

As did Einstein, some would say that the universe has an obviously intelligent design, both simple and beautiful. How could all of this just happen without an intelligent designer? Others would disagree, claiming that all the wonders happening in the universe do so unassisted by the hand of a supreme being. I say[59] the universe is the most marvelous and ingenious structure ever created. It is forever expanding, constantly creating new matter, new galaxies, and new worlds. Who could have imagined it? The universe makes the creation of all physical life possible.

Einstein professed a deep faith that the principles of the universe would be both simple and beautiful, and they are. As stated in the first chapter, "I will present the evidence, you draw the conclusions."

References

1   Today in Science History;   Sir Isaac Newton
     http://www.todayinsci.com/N/Newton_Isaac/NewtonIsaac
     Quotations.htm

2   Genius Among Geniuses http://www.pbs.org/wgbh/nova/einstein/

3   Relativity – by Albert Einstein Chapter XXXII The Structure of
     Space According to the General Theory of Relativity   pg 113

4   The Newtonian Gravitational Constant: Recent Measurements and
     Related Studies - by George T Gillies Department of
     Mechanical, Aerospace and Nuclear Engineering, University of
     Virginia, Charlottesville, Virginia, 22903, USA
     http://www.iop.org/EJ/abstract/0034-4885/60/2/001

5   The Gravitational Constant – by Robert Kritzer
     http://www.physik.uni-wuerzburg.de/~rkritzer/grav.pdf

6   The NIST Reference on Constants, Units, and Uncertainty
     CODATA Internationally recommended values of the
     Fundamental Physical Constants
     http://physics.nist.gov/cuu/Constants/index.html

7   NASA – Jet Propulsion Laboratory California Institute of
     Technology   http://ssd.jpl.nasa.gov/?constants

8   Relativity – by Albert Einstein Chapter XXXII   The Structure of
     Space According to the General Theory of Relativity   pg 113

9   Apollo 15 Hammer-Feather Drop Joe Allen, NASA SP-289,
     Apollo 15 Preliminary Science Report, Summary of Scientific
     Results, p. 2-11
     http://nssdc.gsfc.nasa.gov/planetary/lunar/apollo_15_feather_drop.
     html

10   American Journal of Physics, Volume 56, Issue 5, pp. 413-415
     (1988) Abstract:
     http://adsabs.harvard.edu/abs/1988AmJPh..56..413W

11   Relativity by Albert Einstein;   Appendix IV   pg 133

12   Relativity by Albert Einstein;   Appendix IV   pg 133-134

13   Study of the anomalous acceleration of Pioneer 10 and 11.
            John D. Anderson *et al*
            http://arxiv.org/PS_cache/gr-qc/pdf/0104/0104064v5.pdf

14   Discover magazine;   October 2003 Cover story Nailing Down
            Gravity.

15   Finding the origin of the Pioneer anomaly Michael Martin Nieto &
            Slava G. Turyshev.
            http://www.iop.org/EJ/abstract/0264-9381/21/17/001/

16   The Pioneer anomaly as acceleration of the clocks
            Antonia F. Ranada   http://arxiv.org/abs/gr-gc/0410084

17   The "Pioneer effect" as a manifestation of the cosmic expansion in
            the solar system. J.L. Rosales and J.L. Sanchez-Gomez
            http://arxiv.org/abs/gr-qc/9810085v3

18   A Force to Reckon With   by Alexander Hellemans
            http://www.sciam.com/article.cfm?chanID=sa004&articleID=000
            BB6BE-A7BA-1330-A54583414B7F0000

19   What is Ockham's Razor?
            http://phyun5.ucr.edu/~wudka/Physics7/Notes_www/node10.html

20   Hubble Reveals "Backwards" Spiral Galaxy
            http://hubblesite.org/newscenter/archive/releases/2002/03/image/a

21   Spiral Galaxy NGC 4622 Spins "Backwards"
            http://hubblesite.org/gallery/album/galaxy_collection/spiral_/pr20
            02003a/titles/true/npp/10/

22   NOVA   Galileo's Battle for the Heavens
            His Big Mistake   by Peter Tyson
            http://www.pbs.org/wgbh/nova/galileo/mistake.html

23   Tidal Misconceptions   by Donald E. Simanek

http://www.lhup.edu/~dsimanek/scenario/tides.htm

[24] Tides and Centrifugal Force   by Paolo Sirtoli
mailto:paolo.sirtoli@gmail.com
http://www.vialattea.net/maree/eng/index.htm

[25] A Dynamical Picture of Oceanic Tides by Eugene I. Butikov
Department of Physics, St. Petersburg State University, St.
Petersburg, Russia
http://faculty.ifmo.ru/butikov/Projects/tides1.pdf

[26] infoplease encyclopedia
The Magnitude and Effects of Tidal Ranges
http://www.infoplease.com/ce6/sci/A0861552.html

[27] Moon Tides - How The Moon Affects Ocean Tides
http://home.hiwaay.net/~krcool/Astro/moon/moontides

[28] A Dynamical Picture of Oceanic Tides by Eugene I. Butikov
Department of Physics, St. Petersburg State University,
St. Petersburg, Russia
http://faculty.ifmo.ru/butikov/Projects/tides1.pdf

[29] Tidal Forces and their Effects in the Solar System
by Richard McDonald
http://www.themcdonalds.net/richard/astro/papers/602-tides-
web.pdf

[30] Anomalous Precessions
http://www.mathpages.com/rr/s6-02/6-02.htm

[31] Planetary Science – Mercury Fact Sheet
http://nssdc.gsfc.nasa.gov/planetary/planets/mercurypage.html

[32] The End of Cosmology; Scientific American – March 2008 By
Lawrence M. Krauss and Robert J. Scherrer

[33] Georges-Henri Lemaitre
http://www.mlahanas.de/Physics/Bios/GeorgesHenriLemaitre.html

[34]   Hubble Completes Eight-Year Effort to Measure Expanding
         Universe
         http://hubblesite.org/newscenter/archive/releases/1999/19

[35]   Universe Today website: Chandra Confirms the Hubble Constant
         http://www.universetoday.com/2006/08/08/chandra-confirms-the-
         hubble-constant/ and Spaceflight Now website
         http://www.spaceflightnow.com/news/n0608/08hubbleconstant/

[36]   Dark Matter, Dark Energy: The Dark Side of the Universe;
         Lecture 1 – Fundamental Building Blocks – Sean Carroll
         The Great Courses

[37]   Dark Forces at Work by David Appell;
         Scientific American - May 2008

[38]   God's Equation by Amir D. Aczel
         Chapter 11 – Cosmological Considerations pages 158-159

[39]   The Universe's Invisible Hand; by Christopher J. Conselice
         Scientific American – February 2007

[40]   Density of Outer Space The Physics Factbook
         Edited by Glenn Elert – Written by his students
         http://hypertextbook.com/facts/2000/DaWeiCai.shtml

[41]   The Universe;   Mass, Size, and Density of the Universe Webpage
         from The University of Massachusetts
         http://www.cs.umass.edu/~immerman/stanford/universe.html

[42]   Discover Magazine   December 2005,
         article The Nearby Universe;   page 31

[43]   Milky Way's Age Narrowed Down   by Robert Roy Britt
         http://www.space.com/scienceastronomy/Earth_age_040817.html

[44]   ibid

[45]   PBS Home Video   Stephens Hawking's Universe Volume 1

[46]   The Lighter Side of Gravity;   Chapter 11

by Jayant V. Narlikar

47   New Image of Infant Universe; Goddard Space Flight Center
      http://www.nasa.gov/centers/goddard/news/topstory/2003/0206ma
      presults.html

48   NewScientist   Swift measures distance to gamma-ray bursts
      http://www.newscientist.com/article.ns?id=dn7237

49   Richard P. Feynman; Six Easy Pieces   Perseus Books,   page 117

50   Planck's Constant; The NIST Reference on Constants
      http://physics.nist.gov/cgi-bin/cuu/Value?h

51   Richard P. Feynman; Six Easy Pieces Perseus Books,   page 117

52   Inverse of Fine-structure constant; The NIST Reference of
      Constants   http://physics.nist.gov/cgi-
      bin/cuu/Value?alphinv|search_for=fine-structure

53   Richard P. Feynman;   QED – The Strange Theory of Light and
      Matter Princeton University Press, page 129

54   ibid

55   Fundamental Physical Constants
      The NIST Reference on Constants, Units, and Uncertainty
      http://physics.nist.gov/cuu/Constants

56   Gravity – by George Gamow;
      Chapter 10 – Unsolved Problems of Gravity; page 138

57   Group of History and Theory of Science
      Cesar Lattes and 50 years of the pi meson
      http:///www.ifi.unicamp.br/~ghtc/meson-e.htm

58   John Archibald Wheeler
      PBS Video   The Creation of the Universe

59   Len Kurzawa: lenkurzawa@gmail.com

www.ingramcontent.com/pod-product-compliance
Lightning Source LLC
Chambersburg PA
CBHW031817170526
45157CB00001B/89